The Nature of Visual Illusion

Mark Fineman

Southern Connecticut State College

DOVER PUBLICATIONS, INC.

Mineola, New York

Published in Canada by General Publishing Company, Ltd., 30 Lesmill Road, Don Mills, Toronto, Ontario.
Published in the United Kingdom by Constable and Company, Ltd., 3 The Lanchesters, 162–164 Fulham Palace Road, London W6 9ER.

Bibliographical Note

This Dover edition, first published in 1996, is an unabridged republication of the work first published by Oxford University Press, New York, 1981, under the title *The Inquisitive Eye*. The four pages of color illustrations, following pages 10 and 146 in the original edition, are here reproduced in black and white in their original locations and in color on a new insert following page 20.

Library of Congress Cataloging-in-Publication Data

Fineman, Mark B.
    [Inquisitive eye]
    The nature of visual illusion / Mark Fineman.
        p.   cm.
    Originally published: The inquisitive eye. New York : Oxford University Press, 1981.
    Includes bibliographical references and index.
    ISBN 0-486-29105-7 (pbk.)
    1. Visual perception.   2. Visual perception—Problems; exercises, etc.
I. Title.
BF241.F56   1996
152.14—dc20   96-14001
                                                  CIP

Manufactured in the United States of America
Dover Publications, Inc., 31 East 2nd Street, Mineola, N.Y. 11501

Preface

This book was inspired by my attempts to teach visual perception over the last ten years. I have found that it is much easier to understand a principle when that principle can be seen in action, and so I have emphasized illustrations that people can actually pick up and use—some are even intended to be cut out from the book.

I have also tried to create a book that would be intrinsically interesting, scientifically valid, and readable. Toward that end several of my interests outside of experimental psychology (particularly art and photography) have been incorporated into many of the demonstrations and discussions.

Who might find this book of value? Of course students of visual perception should find material of interest here, but so too could artists, architects, and designers—anyone, for that matter, who is fascinated by the way we see.

I would like to thank my wife Susan for her encouragement of this task, Marcus Boggs and Nancy Amy of Oxford University Press, and Dr. Lloyd Kaufman of New York University for his many helpful suggestions. This book is dedicated to the memory of my father, Morris Fineman.

New Haven   M.F.
January 1981

Contents

The Nature of
Visual Illusion

1. A Light and Color Primer

Almost everyone has seen color samples in a paint store, those small squares of color arranged in neat progessions on cards. A paint manufacturer provides his customers with several hundred samples at most, but imagine for a moment that someone set out to create every possible color that the human eye could distinguish. How many colors would there be? Thousands? Tens of thousands? Actually, the number of discriminably different colors has been estimated to be about 7.5 *million!*

How is that possible? Are there over seven million different kinds of receptors in the eye? Is light itself made up of millions of colors? Color vision is a good topic with which to start an examination of visual perception because it illustrates many of the complexities peculiar to the larger subject of vision. Before we try to understand the workings of color vision, however, it would be a good idea to consider a few fundamentals of vision.

VISUAL PERCEPTION

In the simplest of definitions, visual perception can be reduced to three events: 1. the presence of light, 2. an image being formed on the retina, and 3. an impulse transmitted to the brain.

1. Vision requires a stimulus. This stimulus is normally in a form of energy called *light*. Although sometimes a person claims

to see in the absence of light, as in a dream or hallucination, these are special instances that do not pertain to the immediate discussion.

2. The light stimulus enters the eye, where it is refracted (bent) in such a way that an image is formed on the interior of the eye, a light sensitive surface known as the *retina*. The image-forming

The eye is a light-tight chamber except for the structures through which light enters. The cornea and lens refract the incoming light so as to form an image on the photosensitive interior layer of the eye, the retina. The pupil is a variable aperture whose diameter is regulated by the surrounding iris in response to the level of light in the environment.

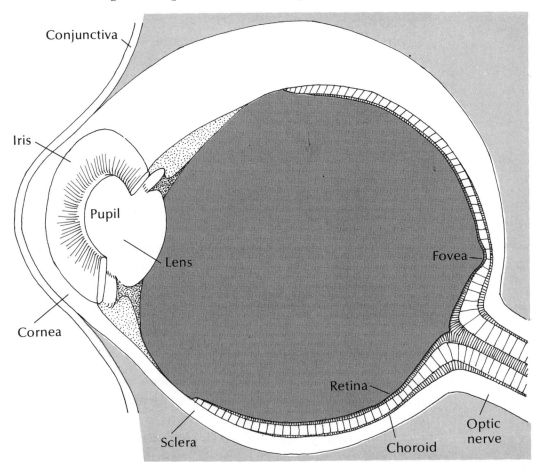

parts of the eye are the *cornea*, a clear bulge at the front of the eye, and the *lens*, located within the eye a short distance behind the cornea. You can see someone else's cornea by asking that person to stare straight ahead while you observe his eye from the side.

The *iris* and *pupil* mechanism, interposed between cornea and lens, regulates the overall level of light that enters the eye. In bright light the pupil constricts, and in dim light it dilates. The pupil itself is an aperture whose diameter is controlled by the surrounding colored iris. The lens and the retina are on the interior of the eye and cannot be seen just by looking at the exterior of another person's eye.

The retina is composed of millions of specialized receptor cells, as well as other types of cells that support the transmission from the retina to the brain. The receptors respond to the light striking them (the image) by triggering a chain of chemical reactions. It is this series of chemical changes that is transmitted from cell to cell, from retina to brain.

3. Thus the third event is the transmission of this chemical impulse to the brain, causing further chemical changes in the billions of cells that constitute the brain. These alterations in brain activity are unquestionably at the core of our responses to light, responses that can be conveniently categorized as "seeing."

To summarize then, any time we talk about visual perception, three events must occur: a stimulus, an image, and the transmission of the impulse to the brain. Because of this, an analysis of any visual phenomenon—including color vision—can be made at one or more of these levels. In this chapter I would like to pay particular attention to some basic features of color vision, while at the same time noting some general features of visual perception.

LIGHT AND COLOR

The visual stimulus, light, is one manifestation of *electromagnetic energy*, which also encompasses such familiar phenomena as X rays, and radio and television transmissions.

Electromagnetic energy can be pictured as having the regular undulations of a wave, one whose distance from peak to peak may

vary. For this reason it is common practice to describe an electromagnetic phenomenon in terms of its wavelength. If electromagnetic energy were arrayed according to wavelength, from shortest
to longest, so as to form a *spectrum,* X rays would occur at the
short end, broadcast bands would be at the long end, and the
waves to which vision responds (*visible light*) would occupy an
intermediate position. In fact, the wavelengths of visible light
vary from about 300 to 700 nanometers. Since a single nanometer
is a mere billionth of a meter long, it can be quickly appreciated
that waves of visible light are still quite short.

In addition, electromagnetic energy (including visible light)
has particle properties. When applied to light, a particle or *quan-
tum* of energy is called a *photon.* This photon is often conjectured
to be like a tiny packet of energy, one that oscillates in waveform
as it travels.

No one is entirely certain why it is that we respond exclusively
to the relatively tiny portion of the electromagnetic spectrum occupied by visible light. Some theorists have speculated that since
these wavelengths are abundant in the sun's radiations, it is reasonable to suppose that we would have evolved to be sensitive to
them. It should be noted that some species of animals actually respond to slightly longer or shorter wavelengths than those of visible light.

Light at different wavelengths within the visible part of the
spectrum varies in color. Longer wavelengths look red and shorter
wavelengths look blue. Surely everyone is familiar with Sir Isaac
Newton's classic experiment: In an otherwise darkened room, a
beam of sunlight was allowed to pass through a slit in a window-
shade. The beam then passed through a glass prism and finally on
to a screen. The prism had the effect of separating out the component wavelengths of the original beam of white light (light
which contained all wavelengths in roughly equal proportions),
creating a spectrum of visible wavelengths. Nature repeats substantially the same experiment whenever a rainbow appears.

Thus we can see a basis for color vision in the stimulus itself,
and one might be inclined to assume that color vision is largely a
matter of detecting various wavelengths of light. Even though this

hypothesis is attractive, it still cannot account for the complexity of color vision. Remember that there are only about 400 wavelengths of visible light, yet we see millions of color tints and shadings. Even if we could discriminate every wavelength of visible light, we could account for the perception of no more than a few hundred colors at best.

INTENSITY AND BRIGHTNESS

The intensity of light is a measure of its energy. It is calculated by multiplying the frequency of light by a constant, named for the eminent German physicist Max Planck who discovered it, and which is therefore called Planck's constant. Now you might suppose that if the energy of light were increased, we would always report that the light appeared brighter. In fact, however, the brightness of an object is only partially related to the energy of the light given off by that object. The intricacy of the intensity-brightness relationship is implicit in the word *brightness* itself, since brightness refers to how people see or respond to the energy dimension rather than to the energy itself. You may find it convenient to think of brightness as a measure of *perceived* intensity. Many researchers consider brightness perception to be an integral part of color vision.

Why has it been necessary to construct this confusing concept of brightness? Why not stick with a straightforward measure of light energy? One reason is shown in the accompanying illustration, in which several gray squares are enclosed by larger squares. In every case the interior squares reflect equal levels of light to the viewer's eyes, a fact that could be easily verified by measuring them with a light meter, or by inspecting the interior squares without the surrounding squares. For example, cut out the white mask and place it over the illustration so that only the inner squares can be seen. When viewed against the uniform white background, the interior squares should appear equally bright. Even though the inner squares are of equal intensity, they nonetheless appear to be of unequal brightness values when the surrounding squares are in view. This contrast effect is attributable

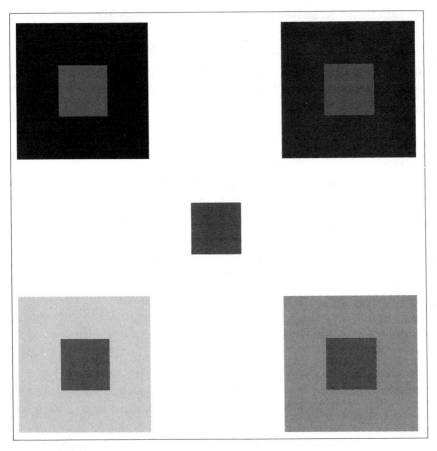

Simultaneous contrast. Even though all of the enclosed squares are of equal intensity, they appear to be of different brightnesses. By cutting out the white mask and placing it over the squares so that only the interior areas show through, you will see that they do indeed reflect equal intensities.

to the viewer's own visual system and will be discussed again in later chapters.

One of the reasons that color vision is so difficult to fathom is that both the nature of light and the nature of the observer's visual system must be considered, as these illustrations may have already suggested. Therefore, let's turn to some of the complications created by the interaction of stimulus and observer.

COLOR MIXING AND COLOR VISION

More than a century ago, Thomas Young, the English physicist and physician, suggested that lights of any three different colors may be combined to create all of the remaining colors of the visible spectrum. For example, if a beam of green light is projected on to a white screen, and a beam of red light is made to overlap the first beam on the screen, the overlapped region will appear yellowish. Thus three lights and their combinations are all that would be needed to create all of the colors of the visible spectrum. In the red and green example, the receptors of the retina are stimulated by wavelengths at two locations on the spectrum, and so this is an example of an *additive color mixture*. (Additive mixtures are typical of the human eye.)

Most people, however, are more familiar with a *subtractive color mixture* such as the type that occurs when paints or pigments are combined. Mixing blue paint with yellow paint yields green. It is called a subtractive mixture because the combination of paints acts to absorb light in certain portions of the spectrum while allowing light at the remaining wavelengths to be reflected to the eye.

The important consideration is that only three colors and their combinations are needed to create all of the spectral colors. Young further theorized that there need be only three types of color receptors in the eye, each responsive to a different portion of the spectrum. Their combined activity could then account for the remaining colors. This *trichromatic theory* of color reception has now been amply supported by the evidence of many laboratory studies, and there is no need to postulate the existence of scores of different cell types in the retina—let alone millions—in order to account for color vision at the level of the eye.

On a more practical level, our knowledge of color mixture and color vision has made color photography possible, as well as color television and color printing. Color film, for example, is a sandwich of three photosensitive layers, each of which reacts chemically to a different portion of the spectrum. The mechanism by which color film operates is therefore analogous to that of the color receptors in our eyes.

COLOR INTERACTIONS

Although the trichromatic theory helped to simplify our understanding of color receptors, it didn't answer all the questions. There are some colors that do not appear in the spectrum arrayed by Newton's prism. Where do colors like pink and brown fit in? Metallic colors (such as silver and bronze) and shades of gray also bear no obvious relationship to the spectrum.

Part of the answer lies in the fact that colored light may be combined with white light to varying degrees. Red light combined with white light will appear lighter, more "pinkish." The grays are a function of intensity of white light and can only be seen as surfaces, not in lights.

Colors can also interact in many ways. One way to see a color interaction is to repeat the earlier contrast demonstration substituting colored paper squares for the gray patches. Any kind of uniformly colored paper will do. Cut out several squares of the same color (a pale green works well) and see how the squares compare when placed against patches of other colors. You should be able to detect subtle differences in brightness, as before, but also slight differences in tint as well.

COLOR CONSTANCY

Understanding color vision is also complicated by a characteristic of observers called *hue constancy* or *color constancy*. So far we have assumed that the light ordinarily illuminating our environment is white light, containing roughly equal proportions of light from all of the visible wavelengths. In actuality white light is the exception rather than the rule. Sunlight only approaches the white standard around noontime. At other times of the day its spectral composition is more varied due to the filtering properties of the atmosphere; sunlight passes through varying thicknesses of atmosphere at different times of the day, and the color of sunlight may

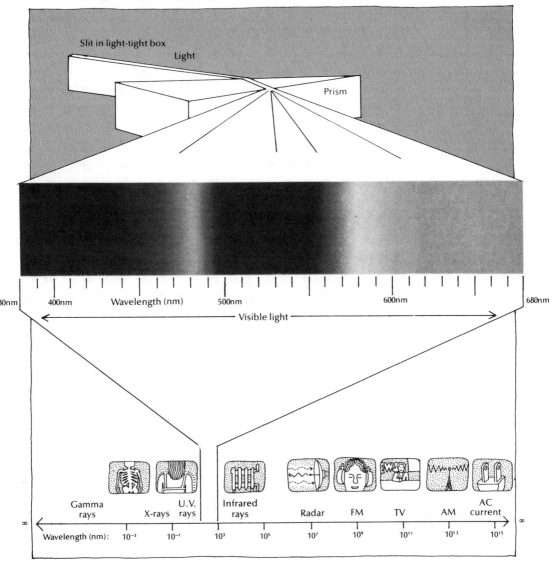

Slit in light-tight box

Light

Prism

480nm   400nm   Wavelength (nm)   500nm   600nm   680nm

Visible light

Gamma rays   X-rays   U.V. rays   Infrared rays   Radar   FM   TV   AM   AC current

Wavelength (nm): 10^{-3} 10^{-1} 10^{3} 10^{5} 10^{7} 10^{9} 10^{11} 10^{13} 10^{15}

The electromagnetic spectrum arranged by wavelength. The relatively small portion of the spectrum that constitutes visible light has been enlarged for clarity. Also shown are some common electromagnetic phenomena and the colors that correspond to positions along the visible spectrum.

(This page also appears in color following page 20.)

An additive color mixture. Prop up this illustration and view it from across the room. The red and green dots will blur sufficiently so that both will stimulate the same receptors.

Color slide film does not maintain color constancy, as our visual systems do. Changes in color balance will be evidenced in slides taken at different times of day and under different sources of illumination.

(This page also appears in color following page 20.)

be affected by dust particles and pollutants suspended in the air. Toward sunrise and sunset light has relatively more energy at the red end of the spectrum and somewhat less at shorter wavelengths than is typical at high noon.

Most lightbulbs also emit a spectrum that is biased from true white light. Incandescent bulbs are reddish, while fluorescent tubes are deficient in long wavelengths and therefore emit a spectrum that is bluish when compared with white light.

We are not normally aware of these discrepancies from white light. Even though objects reflect varying percentages of the visible wavelengths under different conditions of illumination, color perception remains constant. Common objects look much the same whether seen by sunlight, lightbulb, or fluorescent tube. The visual system seems to disregard minor discrepancies in spectral composition so that colors appear much as they would under white light illumination. In this way we are able to maintain a stable world of color under changing conditions of illumination.

One way to demonstrate the constancy of color vision is to take color photographs of the same scene under different conditions of illumination. The best film to use is color slide film, such as Kodak's Kodachrome or Ektachrome, both of which are said to be "balanced" for sunlight at noon. Unlike the visual system of humans, color film cannot adapt to shifts in the coloration of the illuminating light, and its appearance will change as the color quality of the light changes. Try making the following comparison: Take a series of pictures of the same subject at regular intervals throughout the day. You will find that the presence of a few normally white objects in the scene will help to evaluate the results. You may be surprised to see dramatic color changes from picture to picture, changes you had not even been aware of while taking the pictures. Next, take some pictures indoors with the same film, using the light of electric fixtures as the sole illuminant. What would you predict about the appearance of color in pictures that had been illuminated by lightbulbs, as compared with fluorescent illumination?

The adaptability of color vision has its limits. While wide fluctuations in color quality are tolerated in vision, too narrow a band of wavelengths is unacceptable for normal perception of color. To prove this point, obtain a good quality red or green lightbulb and try to identify the colors in magazine illustrations using the bulb as the sole source of illumination. This is a little tricky since you will have had some acquaintance with the colors of many of the objects depicted in the magazine. If you remain perfectly objective, however, you will see that the magazine looks dramatically different under the uniform illumination. If you use a red bulb, the blue printed areas of the page will appear dark. Because much of the light emitted by the bulb is in the long end of the visible spectrum, the red dyes on the printed page will continue to reflect light to the eye, but the bulb emits few wavelengths at the blue end of the spectrum and so there is nothing for the blue dyes to reflect. Therefore areas that looked blue under normal illumination now appear almost black under red light illumination.

The peculiar appearance of colors under homogeneous color illumination is much the same as happens with certain kinds of street lighting, such as the newer sodium or mercury vapor lamps, which emit light only wtihin a narrow band of wavelengths.

2. The Minimum Case for Vision

On an inky-black June night in the South Pacific, at a time late in the Second World War, the pilot of a lumbering Martin patrol bomber prepared his seaplane for takeoff, a tricky business even in daylight. Because of the nature of the patrol, it would be necessary for the plane to circle after becoming airborne and then pass over a seaplane tender anchored below in order for its flight instruments to be calibrated. The disquieting events that followed the takeoff run are best described in the pilot's own words:

> When I turned back to the instruments, I discovered to my horror that the little plane on the artificial horizon was sliding down through the horizon line in a bank of about 40 degrees and at the same time the rate-of-climb indicator was dropping at 200 feet per minute down!
>
> Instantly, I turned the yoke to straighten the wings for level flight and yanked the column back to bring the rate-of-climb indicator up to zero. Then, somewhat shaken, I resumed our gradual turn to 200 feet per minute up . . .
>
> It was a classic case of vertigo. Without any outside reference, I had instinctively followed my sensory impressions and had almost put us in a graveyard spiral. The experience drove home, as nothing else could, the rule to "Trust your instruments, not your senses" when blind flying. Many years later, whenever I thought of it, I broke out in a cold sweat.

Pilots had long known that flying at night or in clouds was potentially dangerous, particularly in the absence of landmarks on the ground. In the pioneering years of aviation, before the development of sensitive flight instruments, it was not uncommon for a pilot to emerge from a cloudbank upside-down and not even recognize the error until the surface of the earth appeared overhead. These disorientations were dubbed *vertigo,* and have been studied intensively over the years. Vertigo has both a visual and a vestibular, or inner-ear, component and either—or both—may give rise to the disorientation.

THE GANZFELD

The visual problem, which first appeared in the literature of perception in the 1930s, can be reduced to this: what is the smallest amount of stimulation necessary in order for vision to occur? Is diffuse, unpatterned light sufficient, or must some additional pattern be present in the field of vision before someone will say, "I see"? As early as 1930, W. Metzger had related the results of an experiment in which observers had been asked to stare at a uniformly lit, whitewashed wall at a distance of 1.25 meters. According to Metzger, "under these conditions, the observer will feel himself swimming in a mist of light which becomes more condensed at an indefinite distance." A new word, *Ganzfeld,* was coined to describe the homogeneous stimulation that had produced the ghostly sensations. Although Ganzfeld can be translated into English as a whole or uniform field, by custom it is left in its original form. The conditions of Metzger's Ganzfeld appear to be related to the atmospheric conditions seen by pilots suffering vertigo.

A Ganzfeld is extremely easy to produce. Although a variety of laboratory apparatuses have been devised for this purpose, a pair of ordinary, white, plastic coffee spoons are a perfectly acceptable substitute. To see the effects of the Ganzfeld, seat yourself comfortably in front of a diffuse light source, such as a large window, or, if practical, the cloudless sky. Then cup the spoons over your opened eyes and note the effects. Do you have the sensation of vision?

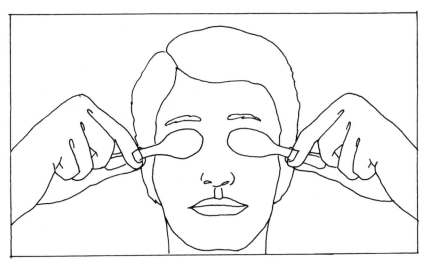

This is the simplest way of producing a Ganzfeld: hold plastic coffee spoons over the eyes. By using colored spoons, you can see the effect of color on perception of the Ganzfeld.

A more permanent viewing apparatus can be made by attaching halves of table tennis balls together with lengths of elastic cord. Care must be taken not to leave seams, printing, or ends of the cord in view in the completed goggles.

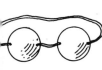

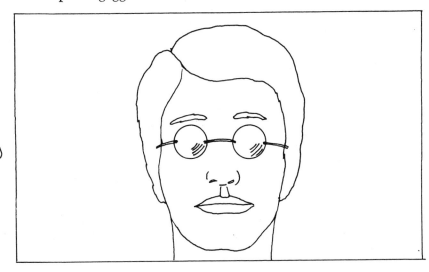

A more permanent arrangement was created by Julian Hoch-berg and fellow researchers in connection with an investigation concerned with the effects of color on viewing the Ganzfeld, a subject to be considered presently. Their contrivance consisted of two halves of a table tennis ball worn over the eyes. Although these diffusing hemispheres were glued to subjects' faces with sur-gical adhesive (a practice I emphatically do *not* recommend!), you might try stringing the Ping-Pong ball halves together with thin elastic cord in order to produce a pair of diffusing goggles. Be sure to cut the ball in such a way that the seam or printing on the ball cannot be viewed. The effects of the Ganzfeld will be ap-parent to you with either the coffee spoons or the table tennis ball goggles.

VISION AT A MINIMUM

What can the Ganzfeld tell us about the minimum conditions of vision? A number of studies leave no doubt that this type of stimulation produces measurable effects, some having to do with changes in the functioning of the eye, and other effects best de-scribed in terms of experience, both of which contribute to the disorientations typical of aviation vertigo. Among the experiential effects are frequent accounts of a visual world variously called: a sea of light, a cloud, a mist, or a fog. Many of these investigations claim that observers of the Ganzfeld see something, no matter how amorphous that something may be. It was just such observa-tions that led the renowned perceptual psychologist Kurt Koffka to conclude that the Ganzfeld was indeed the minimum condition for vision: "we are tempted to say that absolutely homogeneous stimulation causes a minimum event in the nervous system; as little will happen under these conditions as possible."

Koffka, however, had only Metzger's data on which to specu-late, and there is now good reason to believe that Metzger's ap-paratus suffered certain technical faults which were not to be cor-rected until some years later. With improved instrumentation came results claiming that with prolonged viewing of the Ganzfeld, many observers experienced a complete cessation of vision—blindness. Here is a typical description of such a blankout period: "Foggy whiteness, everything blacks out, returns, goes. I feel blind. I'm not even seeing blackness. This differs from the black when lights are out." In this case it could be argued that the Ganzfeld is not even a minimal stimulus for vision and that Koffka's assumption was not to be supported by the evidence of prolonged viewing.

That minimal stimulation produces odd effects was most dramatically illustrated in sensory deprivation studies. This line of research, begun in the 1950s by Bexton, Heron, and Scott, was an attempt to understand how people would react to long periods of excruciating boredom. Human subjects, most of whom were college students, were paid to do absolutely nothing under conditions of reduced sensory stimulation. Except for brief intervals when they were allowed to eat or go to the bathroom, subjects were required to lie motionless on a comfortable bed while wearing diffusing goggles and padded cuffs around the arms and hands. Except for the monotonous whine of an exhaust fan, the room was soundproof.

Although well paid for their participation, few subjects could endure more than a few days of this treatment. The first of these studies did not make it clear whether subjects experienced visual blackout, but many other visual disturbances were common. An unanticipated finding was that in the absence of external stimulation, many subjects experienced vivid hallucinations. Similar kinds of hallucinations have been experienced by long-distance truckers and pilots, usually at night. This line of evidence led to the notion that the nervous system actually requires a certain amount of stimulation for normal functioning, and in the absence of minimal stimulation the senses seem to provide substitute activity.

There are sufficient differences between sensory deprivation studies and Ganzfeld research to warrant caution in making comparisons. Still, I continue to suspect that uniform stimulation is only a sufficient stimulus to vision for short intervals; it is more probably the case that a differentiated stimulus, such as a simple pattern, is more likely to be the elusive minimum stimulus of vi-

sion. Given sufficient exposure to homogeneous stimulation, the
nervous system eventually comes to operate as though patterns of
light and dark were present in the environment rather than the
uniform field produced by the goggles.

The effects of the Ganzfeld are not limited to observations of
white light, for the same loss of vision can be experienced with
uniform stimulation by colored light. Once you have performed
the basic Ganzfeld demonstration, try repeating it while using a
red or green colored light bulb as the only source of illumination,
or else substitute colored coffee spoons for the white ones of the
previous experiment. After several minutes, the colored light
should look desaturated to some degree; that is, it should look
more grayish. It is entirely possible that the loss of color vision
here is just another instance of the adaptation process described
in Chapter 1 in connection with color constancy, and that this
compensation is a property of the receptor cells of the eye.

Since the eyes adapt to color changes independently of one an-

Color adaptation. Hold a strip of colored cellophane or plastic in front
of one eye while both eyes remain open. After the eye has begun to
adapt, remove the plastic and compare how color looks when viewed
alternately by the adapted eye and the unadapted eye.

other, changes in vision caused by uniform color stimulation can be seen directly by comparing how things look for an adapted eye compared to the unadapted eye. Here's how it is done: First expose one eye alone to strongly colored light for a few minutes. It doesn't matter very much if the colored light is patterned or homogeneous, so you can use either a colored coffee spoon or a strip of colored cellophane placed over the eye. The uncovered eye can remain opened during the adaptation period. After exposure is complete, look around you, alternating vision between the two eyes. The appearance of white objects is most obviously influenced by the pre-exposure to a colored field. The reduced sensitivity to the color of the adapting eye is akin to the loss of color vision one finds with long exposures to a colored Ganzfeld.

3. Completion:

Phantoms of the Visual System

Vision is frequently described in machinelike terms, with the eye portrayed as a kind of camera and the visual nervous system a computer that "processes visual information." It isn't wrong to describe visual perception in purely physical terms, but it *is* wrong to make an analogy that confuses certain technological objects (cameras and computers) with our sensory physiology. An analogy isn't an explanation in and of itself. Comparisons of this sort are misleading not only because eyes aren't cameras and brains are not electronic calculators, but also because they ignore a great deal of the complexity and uniqueness of visual perception.

Among the most widely recognized of these distinctive visual characteristics is a phenomenon called *completion*. Completion means that we tend to regard as whole a variety of objects (or their images within the eye) in which portions are missing. In this sense we might say that vision "completes" the stimulus, an indirect way of acknowledging that a specific underlying biological mechanism remains unknown. Nowhere is the principle of completion better shown than in the classic demonstration of the blind spot.

The blind spot is exactly that—a portion of the retina in which there are no receptors. The blind spot occurs because of a curious set of circumstances. The receptor cells of the retina connect with a second class of cells, the bipolars, which in turn synapse with a

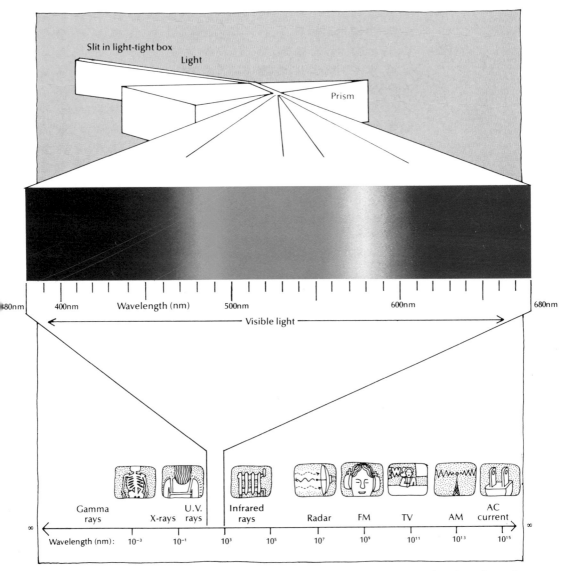

The electromagnetic spectrum arranged by wavelength. The relatively small portion of the spectrum that constitutes visible light has been enlarged for clarity. Also shown are some common electromagnetic phenomena and the colors that correspond to positions along the visible spectrum.

An additive color mixture. Prop up this illustration and view it from across the room. The red and green dots will blur sufficiently so that both will stimulate the same receptors.

Color slide film does not maintain color constancy, as our visual systems do. Changes in color balance will be evidenced in slides taken at different times of day and under different sources of illumination.

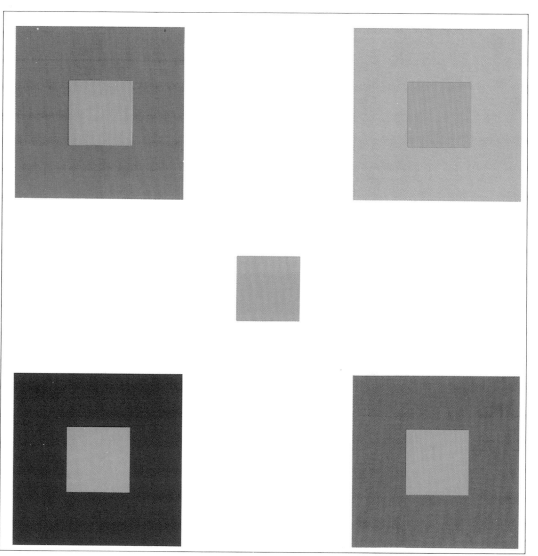

Simultaneous color contrast. The greenish squares are physically identical but they appear to be slightly different from one another when viewed on different backgrounds.

The Bezold effect. Now the green squares do not contrast with the surrounding color but seem to assimilate it; the squares look darker on the dark background and lighter on the light background.

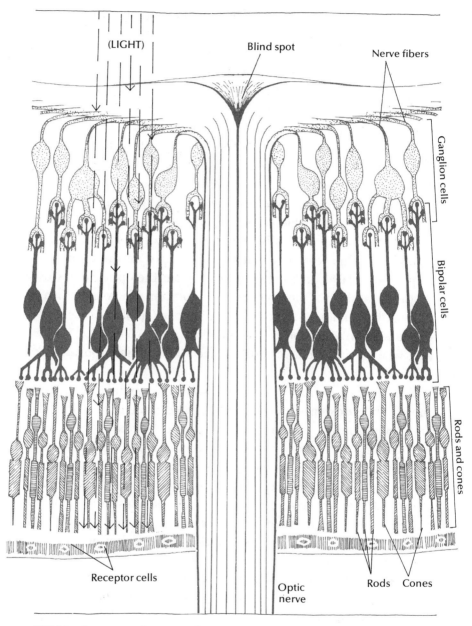

Within the retina there are three main types of cells: receptors, bipolar cells, and ganglion cells. The receptors can be further divided into two types, rods and cones. The fovea (not shown here), a small area in the center of the retina, contains only cone cells. The arrangement of the retinal cells, shown in highly simplified form in this drawing, is responsible for the existence of the blind spot.

third type, the ganglion cells. The elongated axon portions of the ganglion cells collect into a bundlelike arrangement, called the optic nerve, which then exits the eye at a point to one side of the eye. Now one might suppose that these retinal cells are arranged in such a fashion that the light sensitive receptors are toward the interior of the eyeball with subsequent layers behind, nearer the outer surface of the eye. The accompanying illustration shows exactly the reverse to be true: the receptors lie toward the outside of the eyeball. As a consequence of this odd bit of anatomical architecture, it is an impossibility for there to be receptors in the area where the optic nerve exits the eye; there must be a portion of the retina which is actually blind.

Still, we are not normally aware of the existence of the blind spot. Why? One reason is that the blind spot occurs on noncorresponding places on the two retinas, since the optic nerves exit on the nasal sides of the eyes. Even if a portion of one eye's image falls on the blind spot, the corresponding portion of the other eye's image will strike receptors. You may then ask, "But if I close one eye, won't I then be aware of the fact that a blind spot exists?" Probably not.

You are not normally aware of your own blind spots, even though their existence can be demonstrated quite easily. Hold the first demonstration pattern about 10 inches in front of your right eye, keeping the left eye closed. The cross should remain directly in front of the opened eye with the small spot to the right (don't glance back and forth between the cross and the dot). Continue staring at the cross as you slowly move the illustration toward your eye; at a point about five inches from the eye the dot will disappear. To repeat the demonstration for your left eye, just invert the drawing.

Notice that when the dot disappears it is not replaced by a patch of black. The area surrounding the dot just seems to extend over the location where the dot had been. This vivid example of completion is by no means limited to a black figure on a white surround. If you repeat the demonstration using the second pattern, in which black and white are now reversed, the white spot imaged on the blind spot will appear to be completed by the black sur-

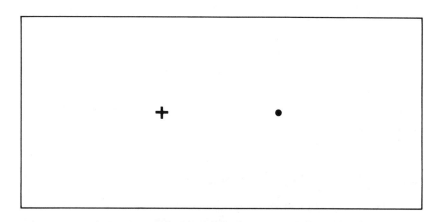

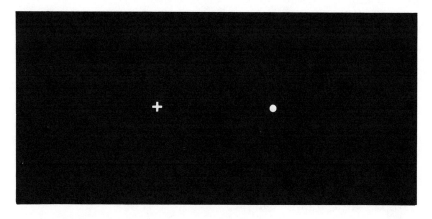

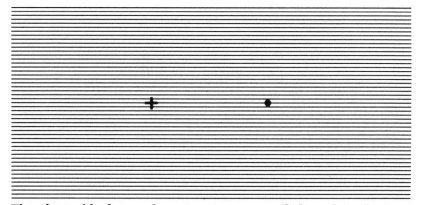

The classic blind spot demonstration is provided in three forms in order to show the plasticity of completion. See the text for complete instructions.

round. Most remarkable of all is the third variation in which the striped background completes the missing figure.

Completion of images formed on the blind spot is a sensory capacity clearly related to the structure of the eye, although we don't completely understand whether the action of retinal cells alone accounts for the phenomenon. As we shall see, other examples of completion cannot be localized so readily.

HOW DOES COMPLETION WORK?

Some commentators have suggested that the prevalence of completion may have something to do with the way in which vision evolved, although no one can say with complete certainty. A variety of evidence has shown that we sample the external world discontinuously rather than in a smooth, coordinated fashion—more like snapshots in an album than the connected frames of a motion picture. Not every external event within the range of vision is imaged on the retina, and not every retinal image is encoded to the brain. The advantage of such a system is that it is extremely economical, permitting us to act in the absence of a complete stimulus; but at the same time it requires a certain amount of fabrication by the central nervous system in order to fill in the gaps between stimulus samples.

PHANTOM CONTOURS

The next three displays were selected to emphasize the point that "filling in" is a widespread property of perception, and also because they are fascinating examples of the principle.

The circular pattern shown on the facing page was devised by Robert Sekuler and Paul Tynan to demonstrate in a simple fashion a phantom contour effect they had discovered some years earlier using sophisticated laboratory equipment. Cut out the disk and punch a hole in the center, then place it on a record turntable set to rotate at 33 revolutions per minute. As the disk turns, you should be able to discern a faint pattern in the black ring, a phantom that appears to be a continuation of the surrounding gray and white radiations. Sekuler and Tynan also found that the appear-

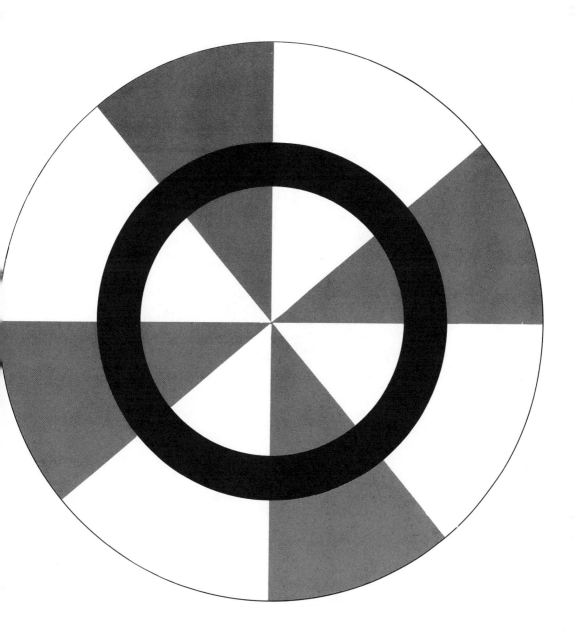

Completion effects are also evident in this disk when it is rotated. The gray and white sections will appear to continue through the black ring when the disk is spun.

ance of the phantom contours is enhanced by viewing the rotating disk in very dim light (even squinting works). You might like to experiment with your own patterns to see if the phantom contours persist under other conditions. Try drawing designs with more or fewer radiations, or in other colors.

The remaining illustrations are all based upon a remarkable drawing first described by L. A. Necker in 1832, which for obvious reasons is called the *Necker cube*. Although Necker was not a psychologist (he was interested in the structure of crystals and first drew the cube in that context), he quickly recognized its unusual perceptual property, one which has continued to intrigue visual researchers to this day. The cube is an example of an ambiguous or reversible figure because with continued inspection it

The Necker cube.

An illusory contour figure.

From D. R. Bradley and H. M. Petry, "Organizational determinants of subjective contour," *American Journal of Psychology*, 1977, 90, 253–262. Copyright 1977 by American Journal of Psychology.

appears to be a skeletal drawing of a cube in either of two distinctly different orientations. If you've never seen this one before, just stare at the cube for a period of time, and as you do, it will suddenly appear to shift its orientation. Once you've become familiar with the effect, you should be able to change the apparent orientation of the figure at will.

In and of itself the Necker cube is not an example of completion, but recently two young researchers, Drake Bradley and Heywood Petry, combined the reversible cube with a second phenomenon, *illusory contour*, to produce an offbeat demonstration of filling-in. A brief digression is required at this point to explain illusory contour. The effect was noted as early as 1904 and can be found in drawings that are constructed in such a way that uniform portions seem to stand out as a distinct geometric form even though the outline of the form is incomplete. Most observers describe the form as being slightly brighter than its background

Bradley and Petry's phantom Necker cube illustrates several principles, including completion by subjective contours and the reversibility of ambiguous geometric forms.

From D. R. Bradley and H. M. Petry, "Organizational determinants of subjective contour," *American Journal of Psychology,* 1977, 90, 253–262. Copyright 1977 by American Journal of Psychology.

even though it is not in terms of a physical measurement of the drawing. Observers also agree that the illusory figure seems to stand out in front of other elements in the picture. The next illustration is a good example of illusory contour: a distinct white triangle can be seen "floating" in front of the black shapes which induce it. Illusory contour drawings exemplify the type of completion we've been discussing, since a whole geometric figure is seen even though sections of the stimulus are absent.

Bradley and Petry's contribution was to point out just how flexible completion may be. Their drawing is an incomplete Necker cube, which, depending upon instructions, can be seen in a variety of ways. For example, if you look at the incomplete form as you had in the original Necker cube, a complete cube will be seen, including portions of the nonexistent lines that should connect the corners! The illusory or phantom lines are contributed by the nervous system since they are not physically present in the drawing. The phantom cube can also be reversed, just like the complete cube.

Now try this variation: imagine that the circular areas of the incomplete cube drawing are actually holes in the page and that the cube is suspended *behind* the surface of the page. This arrangement may take a bit longer to achieve, but be patient. When you can see the cube in this fashion, notice that the illusory lines disappear, even though the cube continues to reverse. The disappearance of the illusory contours when the cube is seen "behind" the page is consistent with a theory of illusory contour that attributes the phenomenon to depth perception, i.e., the black-inducing elements of the pictures always suggest that the illusory object is on top of them, never behind. If you now look back at the illusory triangle, the incomplete circles and triangle outline are consistent with the interpretation of a white triangle superimposed on top of the black areas, not behind them.

4. Motion Receptors

There are some ten billion cells in the human brain, each measuring but a few thousandths of an inch in length, and all connected in such labyrinthine ways that the workings of an electronic computer are simple by comparison. It is this combination of extreme complexity and smallness of scale, coupled with the physical inaccessibility of the living brain, that have made it so difficult for us to fully understand the relationships between the brain and our behavior. Even though impressive advances in instrumentation have been made in recent years—techniques that have made it possible to study the living brain directly—our knowledge of this enigmatic organ comes largely from a combination of evidence taken from both physiology and behavior. This interdisciplinary approach has led to surprising insights about the way the nervous system operates; in the case of motion perception there is an excellent correspondence between the physiological and psychological evidence. Both sources strongly suggest that there are cells in the visual nervous system that respond not only to the motion of objects, but respond differentially to different directions of motion.

ILLUSIONS OF MOVEMENT

The behavioral data on the nature of motion perception are rooted in observations concerning some unusual *aftereffects* of watching a repetitiously moving scene. These motion aftereffects are so common that you are likely to have seen them yourself, although you may not have thought much about them at the time. An Englishman, R. Addams, provided in 1834 an early account of one such effect. He noted that when he stared continuously at a waterfall for a brief period of time and then fixated the stationary landscape on shore, the fixed objects appeared to float upward momentarily. Somehow the continuous downward cascade of water had created conditions whose effects had persisted after the moving stimulus was no longer in sight. Addams's *waterfall illusion* bore certain resemblances to features of the *bridge illusion*, in which an observer, leaning out over the guardrail of a foot bridge and observing the water passing beneath him, would subsequently see fixed objects on shore float in the direction opposite to the water's motion. (An additional consequence of the bridge illusion was that the observer would also feel *himself* to be moving in the opposite direction to the water's movement as he stared at the passing stream. Although this eerie *induced self-motion* is important to understanding how we determine our own bodily motion, it is less pertinent to the present discussion.)

A convenient way to see the waterfall illusion is to watch the face of a poorly tuned television set at moderately close range, perhaps a yard or so distant, as the picture repeatedly rolls upward or downward. If you stare at the center of the drifting picture for about a minute, and then quickly shift your gaze to a nearby stationary object, the second object should appear to drift or float for a short interval of time. If the stationary object is finely patterned, like a newspaper page, the pattern will seem to "crawl."

At one time it was thought that motion aftereffects might be caused by eye movements. Perhaps, it was reasoned, the continuous stimulation of the eyes by movement in a single direction caused the eyes to pursue the movement and these pursuit eye movements persisted after the moving stimulus was removed. The theory can be faulted on logical grounds (why must we assume that eye movements cause the appearance of motion in objects when it is equally plausible that the appearance of motion in objects causes eye movements) and on factual grounds. With more

Addams's waterfall illusion was seen by first staring at a waterfall for
a brief interval of time. When the observer's gaze was then shifted,
stationary objects appeared to levitate. The latter is a typical motion
aftereffect.

complex displays of motion, the aftereffects also increase in complexity. So if the moving stimulus consisted of two vertically moving patterns, one up and one down, the aftereffect would be half
down and half up, the reverse of the original movement. As for
the eye movement theory, how could the eyes move in two directions at once!

The inadequacy of the eye movement explanation is emphasized by observing the aftereffect of a revolving spiral, a variation
logically enough called the *spiral aftereffect*. The simple spiral in
this book will produce a dramatic aftereffect—cut it out, punch a

hole in the center of the spiral, and place it on a record turntable set to revolve at 33 rpm. Stare at the moving spiral for a minute or so (it will appear to expand as it revolves), and then stop the turntable abruptly. The now stationary disk will look as if it is constricting, the reverse of the inducing stimulus. Since it is impossible for the eyes to move in ways that correspond to the apparent motion of the stationary spiral, a different sort of theory must be sought. By the way, the aftereffect works just as effectively when the stationary object is a page of type, or even someone else's face, a most peculiar variation of the illusion.

If you would like to experiment with spirals at greater length, there is a technique on page 33 for making your own. Spirals can also be used to show that antagonistic motions produce antagonistic aftereffects. One of the patterns shown here will produce two bands of motion, one seemingly expanding and the other contracting, when rotated on the record turntable. Since the corresponding aftereffects of both motions will be seen simultaneously, we can conclude that the channels by which motion detection work must be independent of one another in terms of direction.

MOTION RECEPTORS

These spiral aftereffects hint at one way the visual nervous system may function. Imagine that somewhere in the eye or brain are cells that respond only to particular directions of object motion, up, down, left-to-right, and so on, one type of cell or assembly of cells for each direction. If such "motion receptors" exist, then it is probably the case that they operate independently of one another, the point just made by the double spiral demonstration, and by many excellent experiments. But it is also the case that they might in some way work in opposing pairs, for only when opposing types of receptors are equally stimulated would an object appear stationary.

Let's consider the waterfall illusion again in terms of an opposing-pair model of motion receptors. Suppose that there are "down receptors" and "up receptors," and when unstimulated,

A simple spiral will appear to expand or contract when rotated on a record turntable. After continued viewing of the moving spiral, a stationary pattern will seem to move in the opposite way.

This combination spiral produces both apparent expansion and contraction when rotated. The aftereffects seen in a stationary pattern are the reverse of those that produced them.

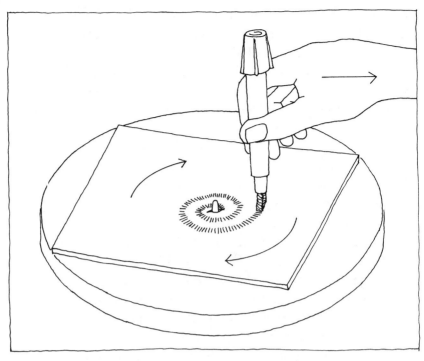

Bill Gray has invented several clever ways to draw your own spirals, one of which is illustrated here. Cut a slit in the center of a square of cardboard, push the cardboard onto the spindle of a record turntable, and then move a brush, pen, or marker from the center of the rotating cardboard toward the edge.

From *Studio Tips for Artists and Graphic Designers*. © 1976 by Litton Educational Publishing, Inc. Reprinted by permission of Van Nostrand Reinhold Company.

both fire spontaneously at a steady rate. When exposed to the waterfall, the viewer's down receptors are made to function at a higher rate than their opposite numbers, and the water is seen to flow downward. Since the down receptors are worked at a higher rate, they are also more likely to become fatigued, so that when the external stimulus abruptly ceases, their rate of operation will be somewhat less than the steady firing of the unfatigued up receptors. Until they recuperate, the down receptors will fire at a lower relative rate than the up receptors, and it is this differential that could theoretically account for the aftereffect. Other models hypothesize the existence of motion detectors but claim that they need not be organized exclusively in opposing pairs.

Although the preceding description is only an educated guess, it is a very good guess in the sense that it adequately accounts for motion aftereffects. The question remains, do motion detectors actually exist or are they just a convention of theory? Conveniently for our purposes, physiological research confirms the existence of motion receptors. A widely used technique in this type of research involves the implantation of fine electrodes into those portions of the nervous system associated with vision (frequently in the optic nerve or the visual cortex of the brain). Because the wire of the electrode is so slender, it can be used to record the activity of a single cell as the organism views its world.

Using this microelectrode recording technique, David H. Hubel and Thorsten N. Wiesel explored the responses of single cells in the brain of lightly anaesthesized cats as simple patterns of light were shone on a screen in front of the animals. They found that there were specific kinds of receptors in the cat's visual cortex, some of which operated only when slits of light were displayed, others which responded only to edges, and still others that responded only to motion. The finding that specific cells responded to one direction of motion but not others was particularly intriguing. It is now widely accepted that motion aftereffects are related to the existence of motion detectors, and that motion receptors are a basic mechanism in the functioning of the brain during motion perception. The particulars of the system remain to be worked out, but even now research is revealing whether the detectors are established at birth, and exactly how motion detectors might function in higher animals.

5. The Illusion of Movement

Interest nowadays in the art of motion pictures is formidable. For many people this attachment has grown beyond a simple infatuation into a passionate love affair. Yet, even as we scrutinize the artistic achievements of the movies in the minutest detail, rarely do we ponder the process itself. By now surely everyone knows that nothing on the movie screen actually moves, that we see a series of still images in rapid succession. Why then does it appear as if things are moving, and so convincingly that the cinematic version cannot be distinguished from real movement? It would be comforting (and convenient) if I could supply an answer in a paragraph or two, but in actuality many pieces of this particular visual puzzle wait to be fitted.

The perceptual principle by which motion pictures work has been known for at least one hundred and fifty years. In psychology the terms *apparent movement* and *illusory movement* are most commonly used to describe instances when successive stationary images of objects are interpreted as real object movement. *Real movement,* on the other hand, occurs only when an object is continuously displaced through space. Notice something here: real movement is distinguished from apparent movement on the basis of how an external stimulus behaves. However, from the observer's point-of-view, motion is motion, and with only rare exception can real movement be distinguished from apparent move-

ment. This point will be of even greater importance in the next
chapter when we consider the so-called wagon wheel effect. For
now just note that in addition to motion pictures, several technological processes, including television and stroboscopic illumination, exist because of apparent movement. We'll look at how
apparent movement works, while in the next chapter I will give
additional examples, and will speculate upon why two modes of
motion perception, real and apparent, exist.

PERSISTENCE OF VISION

In its simplest form, apparent movement can be reduced to two
events; first one appears, then the other, their combined appearance yielding the appearance of movement. Of course a number
of variables, like the brightness of the images, and their form, will
determine if movement will actually be perceived by the observer.

No apparatus whatever is needed to illustrate this simplest
case—your own two eyes will do. Because the eyes are separated
by a distance of about two and a half inches, it follows as a consequence of geometry that the images formed on the two retinas
of those two eyes will differ slightly from one another in most circumstances (this difference is called *retinal disparity*). To see apparent movement at work, hold your finger up about two feet in
front of your eyes while staring at a distant object. Quickly blink
one eye, then the other, paying attention as you do to the appearance of the nearby finger. What you will see is not the rapid juxta-
position of two separate and slightly different images, but a single
image that will appear to move in jerky fashion from side to side.
Somehow, the successive images have been integrated into a fluid
sequence of movement, rather than being seen as independent, alternating events. It must be the case that once an image is formed
on the retina, its effects tend to persist momentarily after the image is removed, and the gaps (the intervals when the eyelids were
closed in the alternating eyes demonstration) between exposures
are interpolated by the tendency of vision to persist. The sequence
of events can be clarified by substituting a primitive shutter for
the blinking eyelids. The shutter is nothing more than a hole

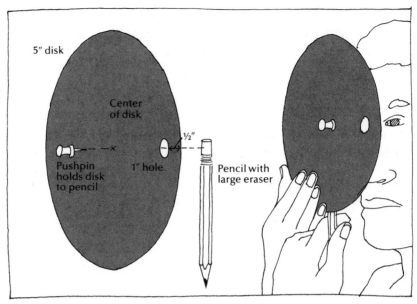

You can make a simple shutter by first cutting out a circular hole along the edge of a cardboard disk. Pin the disk to one end of a pencil using a thumbtack or pushpin. It helps to enlarge the hole slightly so that the disk will turn freely. By making the disk in the dimensions shown, you will make a shutter that exposes the eyes alternately as it spins in front of your face.

punched in a paper disk, which is spun before your eyes, alternately exposing the left eye then the right eye to view, with intervening blank periods. Those readers who have difficulty blinking alternately might do well to skip the earlier demonstration and perform the rotating shutter demonstration alone.

Persistence of vision has long been held to underlie apparent movement. Its existence was first convincingly shown by an English physician, John Ayrton Paris, who popularized a toy that he dubbed the *thaumatrope* around the year 1826. The thaumatrope was little more than a small disk with a picture printed on either side. For example, one side of the disk might portray an empty bird cage and the opposite side a bird. When the disk was twirled by twisting two strings attached to the disk, a single fused image was seen. In this example, the bird appeared inside the cage. The thaumatrope shown here can be manipulated by attaching two strings or by taping the disk to a rod (such as a pencil) and then rotating the rod between your palms. Since the two pictures never come into view simultaneously, it must be assumed that a charac-

To make a thaumatrope, cut out the figures enclosed by the joined circles, fold them over a piece of cardboard, and glue or tape them together. Attach strings to either side of the disk and twist them vigorously. You could also affix the disk to one end of a pencil and roll the pencil back and forth between your palms.

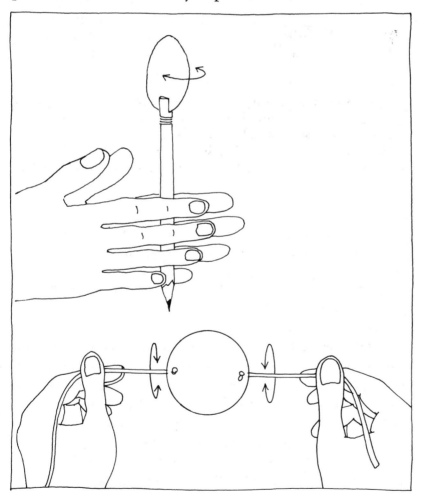

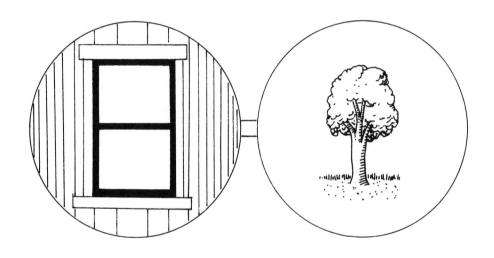

teristic of the observer's visual system accounts for the integrated display seen in the rapidly whirling thaumatrope.

Incidentally, Dr. Paris was more than the inventor of novelties. He had sets of thaumatrope cards printed up that were sold in conjunction with a pun-filled novel. The book went through eight lucrative printings before the good doctor passed away in 1856.

The thaumatrope evoked some comment in scientific circles. However, it was not itself a demonstration of apparent movement, only the principle of persistence of vision. Paris's principle has actually been held in higher esteem by cinema buffs than by students of vision. For as Robert Woodworth pointed out in 1938, persistence of vision can only account for vision during the gaps between exposures of discrete, stationary stimuli. It actually tells us nothing about why two successive stimuli should be seen as a fluid change in spatial position.

THE PHENAKISTOSCOPE

It remained for a contemporary of Dr. Paris, a Belgian scientist and artist named Joseph Plateau, to demonstrate the more complex phenomenon of apparent movement. Plateau's work is often overlooked in contemporary accounts of the history of vision research, which is a pity since he was an enormously skilled and original thinker. By all accounts he was a talented artist who elected to pursue an academic career, eventually earning a doctorate in physical and mathematical sciences from the University of Liège in 1829. In his doctoral thesis Plateau described the workings of a pulleyed machine that he had used to alter the appearance of rotated drawings of his own design. By modifying the principles of the device, Plateau succeeded some three years later in producing a simple demonstration of apparent motion. The second apparatus was a circular, slotted card around whose perimeter were drawn a number of upright figures, each of which differed ever so slightly in posture from its predecessor. When the card was spun and viewed in a mirror, the figure appeared to move in a repetitive dance. Plateau called his invention the *phenakistoscope*, and an example of one is provided here so that you

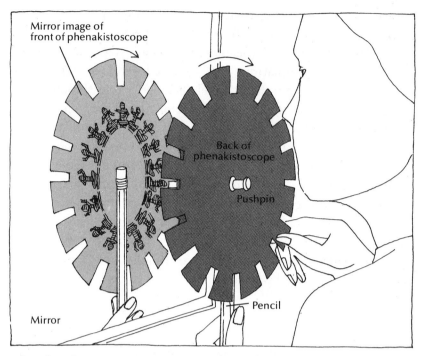

The phenakistoscope illustrates the basic features of apparent motion. The one reproduced here is from the collection of the Yale Center for British Art and was published in 1833 by Ackermann and Co. of London. T. T. Bury, the artist who created it, called the contrivance a *Fantascope*. Carefully cut out the disk, including the slots along its edges, glue on a lightweight cardboard backing, and mount the disk on the eraser end of a pencil in the same way that you mounted the rotating shutter. The wheel should spin freely. While holding the device with the pictures facing a mirror, rotate the disk and view the reflected images through the slots.

Yale Center for British Art, Paul Mellon Collection.

can see how it works. Cut out the card, being certain to remove the narrow slits between the drawings. Then, with a thumbtack, pin the disk to the eraser end of a pencil; enlarge the hole slightly so that the wheel will rotate freely. To operate the phenakistoscope, hold it in front of a mirror with the drawings facing the mirror; as you stare through the slots, spin the wheel and observe the apparent movement of the figures.

As the disk passes your eye it acts as a shutter, intermittently blocking your view between exposures to successive figures. It is just a more sophisticated version of the one-holed rotating shutter you examined earlier, but in the case of the phenakistoscope, a single eye views a succession of stationary drawings rather than the alternating binocular views of the simpler shutter. Motion pictures add another level of ingenuity to Plateau's principle in that a series of still photographs recorded on the film are projected on to the screen, each for a fraction of a second. Between exposures a motor-driven shutter within the projector blocks the image, much as it was done in the phenakistoscope of more than a century ago.

Plateau, in a show of generosity, refused to patent his device so that it might reach as large an audience as possible. Substantially the same contrivance was invented independently by a German geologist and mathematician, Simon Ritter von Stampfer, reports of which appeared in 1834. Stampfer called his motion machine a *stroboscope,* and possibly because he less philanthropically decided to patent his invention, it is *stroboscope* that remains a part of our vocabulary while *phenakistoscope* has faded into obscurity. Plateau continued his researches in the field of vision, even after a disastrous experiment ruined his eyesight (he is said to have stared directly at the sun for some twenty-five minutes). How ironic it must have been that this gifted scientist would never again witness the results of his insights about vision.

6. The Wagon Wheel Effect

It's fairly simple to demonstrate the difference between real movement and apparent movement. It is far more difficult to show why the two systems exist in the first place. Is the distinction between real and apparent movement limited to the stimuli that produce motion perception, or does it extend to the viewer as well? There are other questions that follow: If the distinction is a biological one, how is this feature of the nervous system arranged? What utility could nature have intended for the seeing of two kinds of movement? If there really are two modalities underlying motion perception, how are they related to one another?

The state of our knowledge allows us only tentative answers to these questions. Nevertheless I find it irresistible to speculate based upon the data that exist. In this chapter some apparent movement demonstrations will be presented and the real versus apparent motion distinction will be further explored.

Let's start simply. Apparent motion and real motion don't really look very different from each other. When you look at a movie or TV screen, things actually appear to move. The fact that real movement cannot be easily distinguished from apparent movement has also been recognized in the controlled environment of the laboratory. In several experiments, subjects were required to discriminate between a simple display of apparent movement and a comparable real movement display. Observers were unable to

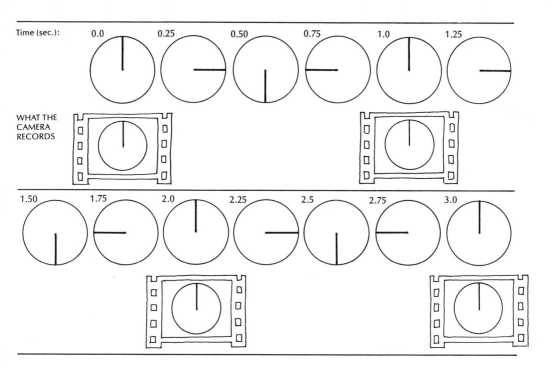

The first series of illustrations shows a simplified one-spoked wheel in motion during an interval of three seconds. A correspondingly simple motion picture camera recording at one-second intervals would repeatedly record the wheel in the same position.

make reliable distinctions between the two displays, even though they saw little more than a single luminous line moving back and forth across their visual fields.

But if we were to examine our own experience more carefully, we would be quick to notice that the most ubiquitous examples of apparent movement (the motion picture and television screens) don't look precisely like the events they portray in the real world. Of course this lack of correspondence occurs for a number of reasons, motion perception being only one of them. For example, there are incongruities of size since motion picture images are often larger-than-life while those of television are miniaturized.

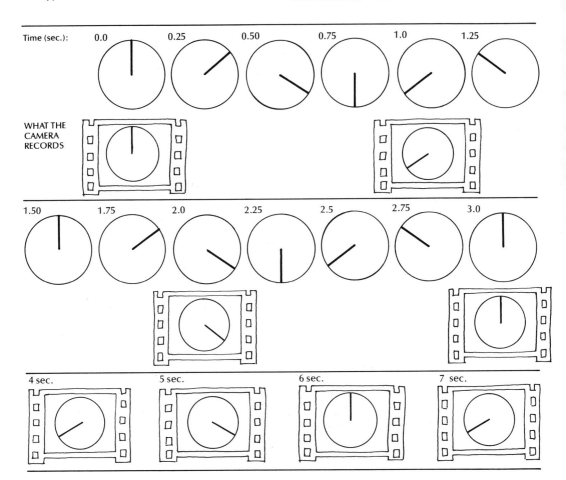

In the next series of illustrations, the same process is repeated for a slower wheel. The samples recorded by the motion picture camera can be cut out and assembled into a flip book. Although the wheel actually rotates in a clockwise direction, the sequence exposed by the camera shows the reverse. The apparent direction of rotation seen in the finished film will thus depend on the rate at which the frames are exposed and the speed of the wheel.

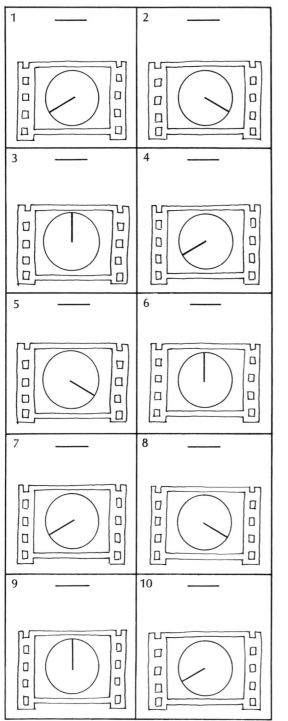

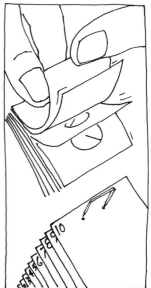

There is also a failure to depict three-dimensional space accurately in both television and motion pictures since both are two-dimensional representations of a three-dimensional world. Both types of display are bounded by the edges of the screen, and both are sometimes monochromatic (black and white) while the real world is variously colored. But if we can bring ourselves to set aside these distinctions, we might agree that the motion of TV and the movies correctly portrays the motion of the real world. There is, however, one curiosity of apparent movement that warrants further examination. Consider the following examples:

In a televised horse opera, the wheels of a stagecoach appear to revolve in a direction that contradicts the motion of the coach itself.

A fighter pilot starts up his aircraft and the movie portrayal shows an intoxicated blur in which the propeller turns first in one direction, then another, accelerating and decelerating, until the engine establishes a constant rate of rotation.

For simplicity's sake, all such instances will be described as occurrences of the *wagon wheel effect*. Although the effect can sometimes be seen in everyday life, it is particularly endemic to apparent movement displays. Perhaps it could be argued that the wagon wheel effect is a reliable way of differentiating real from apparent movement. In explaining the effect, however, I hope to show that such is not the case at all.

It is important to keep in mind the notion that television and cine-cameras are sampling devices; a 16mm-format motion picture camera, for instance, exposes twenty-four frames of film per second. Likewise, the processed film is projected at the same rate. Now let us suppose that we had a low-speed camera, one that exposed film at a single frame per second, and also imagine that it was filming a single-spoked wheel that rotated at exactly one revolution per second. The action of the wheel over an interval of time has been illustrated along with the samples (exposed frames) that the hypothetical camera would record. Even when the wheel turned one complete revolution, the camera recorded only its starting and terminal positions, which are identical, and did so repeatedly as the wheel revolved. When projected, the successive

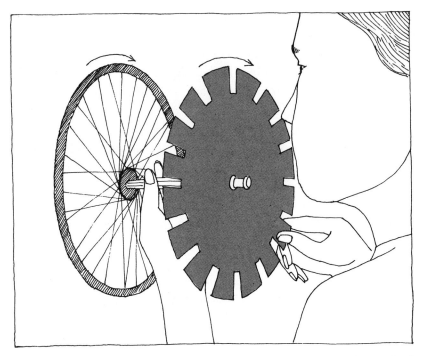

Another demonstration of the wagon wheel effect. Mount the spoked tire on lightweight cardboard and pin it to one end of a pencil with a pushpin or tack. Be certain that it will spin freely. Pin the slotted phenakistoscope disk on the other end of the pencil, blank side toward you. Spin both wheels in the same direction and observe the movement of the spokes through the slits of the rotating phenakistoscope. Even though the tire actually rotates in a single direction at a decreasing rate of speed, it will appear to change speed and direction when viewed through the shutter-like slits of the phenakistoscope disk.

images would depict a stationary wheel. To prove this point, cut out the individual frames and make a flip book as instructed. It should be easy to see that if the wheel turned at a speed slightly slower than one revolution per second, the successive images recorded would be those shown in the next series of illustrations. In this case, the camera—and the flip book—would show the wheel to be rotating in a direction the reverse of its true direction. Whether the revolving wheel will ultimately be seen to be at rest, rotating clockwise, or counterclockwise, and at what rate of rota-

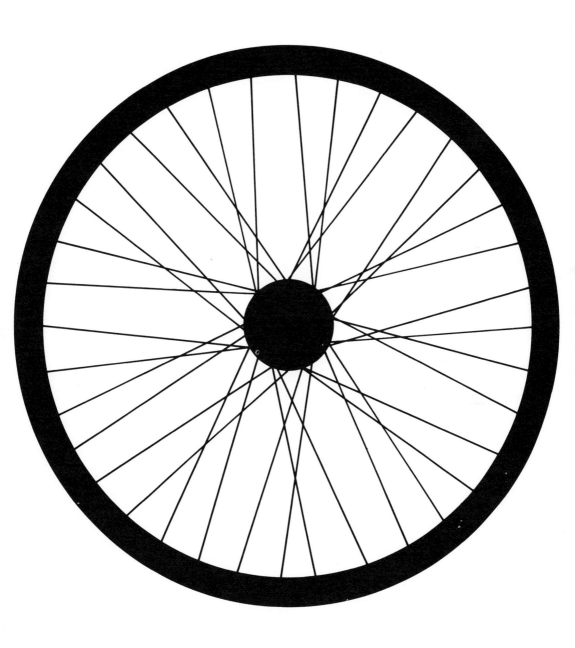

tion, would depend on the wheel's true rate of rotation and the exposure rate of the recording instrument. If the wheel were to uniformly accelerate (as the propeller did in the earlier example), the film depiction would seem to both accelerate and decelerate, and to alternate the direction of rotation.

I've gone into some detail about the wagon wheel effect because it is intrinsically interesting, and also because an understanding of it will reinforce the earlier discussion of apparent movement. It should also be clear by now that the wagon wheel effect does not distinguish real from apparent motion. The effect is attributable to the mechanics of cameras, not to the nature of vision. For even though it may seem incongruous, those crazy wagon wheels persist in looking as if they're moving opposite to the direction of the vehicle to which they are attached—and as far as we the observer can tell, they are!

Some of the earliest comments of a scientific nature on apparent movement were made in reference to the wagon wheel effect. Even though some of these observations predated even the crudest motion picture machines, the effect can occasionally be seen in naturally occurring circumstances. Joseph Plateau was one of those fascinated by the phenomenon, and among the effects produced by a contrivance he invented for his doctoral research was a reconstruction of the wagon wheel effect. By using the slotted phenakistoscope disk of Chapter 5 in conjunction with the spoked tire drawing, you can see what Plateau saw. If you've followed the directions correctly, the spokes of the drawing will not only seem to rotate crazily, but may appear to bend slightly as well.

In summary, then, it can be said that the appearance of movement is not sufficient evidence for determining that something is actually moving, since as we have seen, two kinds of stimuli, one continuous and one intermittent, can produce the same response in the observer.

MOTION DETECTORS

We haven't ruled out the possibility that there are two response systems within the viewer, perhaps two kinds of neural cells, that are affected differently by the two kinds of movement stimuli. It would then be possible for each neurological subsystem to initiate the same overt response in the observer. The pioneering research of Hubel and Wiesel in the 1960s demonstrated that specific cells exist in the visual cortex of the cat's brain that function only in response to moving stimuli in the animal's visual field. Subsequent work has shown that these motion detectors exist in the brains of other species, including the brains of primates. Presumably they exist in the human brain as well. It may be that motion detectors can be grouped according to whether they react to continuously displaced objects in the environment or react to those that obey the rules of apparent movement. Current research has not been directed to this issue, although it is one which is certainly within the scope of established research technology.

Still another possibility is that the same cells of the visual nervous system mediate both kinds of stimuli. This hypothesis is plausible for the following reasons: Even though an external stimulus can be said to be continuously displaced, or intermittent, it is far more difficult to make the distinction for the corresponding images on the retina. The recurrent blinking of the eyelids as they wipe across the front surfaces of the eyes can be thought of as a biological shutter, interrupting the continuous transition of images across the retinas. This interruption of retinal image movement is compounded by the ceaseless movements of the eyes themselves; the eyes make many small excursions, even when we consciously try to fixate our gaze. One consequence of such involuntary eye movements is that during the extremely fast linear movements known as *saccadic eye movements,* the visual threshold is raised, rendering the blurred images useless during the movement. So again the picture emerges of a visual system that synthesizes behavior derived only in part from the fragmentary images of the retinas. Because vision is this way, we must accept the fact that even when an external stimulus moves continuously, its retinal counterpart is more likely than not going to be discontinuous.

Motion perception is therefore another example of the completion phenomenon discussed in Chapter 3. At the retinal level, at

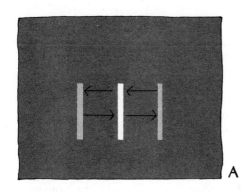

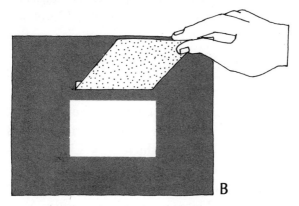

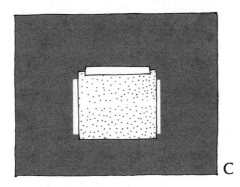

An illuminated vertical bar is made to oscillate back and forth (a). This is real movement. As the speed of oscillation increases, the bar appears to be a solid rectangle of light (b). But if a mask is placed over the middle of the rectangle (so that only the endpoints of the path are exposed to view) the appearance of movement is restored (c). This is apparent movement.

Real movement and apparent movement are a function of relative velocity and may be complementary processes.

least, vision is a matter of fragmentary sampling, and this characteristic provides the best clue for the speculation surrounding the
dichotomous nature of motion perception. In the end it must be
asked, "*Why* should two very different kinds of stimuli give rise to
the same perception of motion?" Surely nature did not engage in
millions of years of evolutionary transformation just to make us
movie critics. It is more probable that flux is the norm in vision
and that our ability to see movement is based upon the conserva-
tive premise that movement which occurs at the retinal level will
be sporadic. When we see movement on the television or motion
picture screen, I suspect that we witness an intentional arrangement of stimuli that triggers the same responses in the eye and
brain that occur in the viewing of all movement.

Finally, let us consider the possible relationship between real
and apparent movement. One provocative line of evidence was
constructed in experimental studies conducted by Lloyd Kaufman, in association with several of his graduate students. Although the precise experimental procedures were too complex to
describe here, basically their subjects were asked to judge a luminous vertical line: 1. when seen as it moved continuously along
a horizontal path, (real movement) and 2. when the middle of the
line's path was obscured, so that only the endpoints remained exposed. (apparent movement).

By carefully recording the rate of the line's displacement, the
researchers found that real object motion was detected at lower
velocities, but at high velocities the continuously displaced line
blurred. Yet at these higher velocities, when only the endpoints of
the line's path were visible, motion perception was restored. Therefore it seems to be the case that the two phenomena are complimentary, and if the two-detector model is correct, then the real
movement detectors may operate at low object velocities and the
apparent movement detectors at relatively higher rates of object
motion. It is also possible that the findings of Kaufman and his
associates can be incorporated into the single detector model, but
only additional research will be able to tell for sure.

7. Motion Perception: Learned or Innate?

We've now made an important distinction between two classes of motion perception, real and illusory. But regardless of type, we must ask whether the ability to perceive motion is inborn or if it must be acquired. The distinction between innately determined behavior and acquired behavior is a legacy given to us by generations of philosophers. Nonetheless, whether sensory behavior is learned, genetically determined, or a combination of both, is not an abstract philosophical dilemma; it remains a central question for visual scientists and has been applied to all of the visual functions, not just motion perception. One reason it is difficult to examine the learned-innate question is that the organisms most often tested (animals or human infants) cannot speak, so it is pointless to ask a one-day-old baby if it can see motion. More subtle techniques, not dependent on verbal reports, are required. In this chapter we'll examine one such nonverbal indicator of motion detection and how it was applied to the issue of whether the ability to see motion is inborn. Instructions are also provided for you to re-create some of the effects discovered by this research.

Several investigators have used as their marker of motion detection a behavior called the *optokinetic reflex*. It can be seen in the eye movements of people whenever they view repetitiously moving objects, like the windshield wipers on an automobile, or railroad cars seen from the platform as a train rolls into the sta-

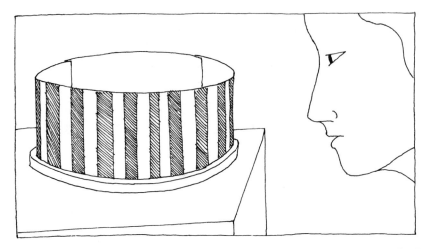

Cut out the strips of striped paper, then insert the tab at the end of one strip into the slot of the next—this will make a striped drum. For the first experiment, keep the stripes on the outside of the drum.

Nystagmus movements. With the stripes on the outside, place the drum on a slowly revolving record turntable and seat an observer so that her eyes are at the same level as the paper drum and about one foot distant from it. As the drum revolves, notice how the observer's eyes flit back and forth in an involuntary reflex.

tion. Typically, the eyes will flick rapidly from side to side in an involuntary response to the movement. This kind of eye movement is known as visual *nystagmus*. In other species the eyes are not so much affected as is the overall behavior of the organism. For example, a fish or insect suspended inside a rotating drum, the interior of which has been painted with alternating black and white stripes, will move in the direction of the drum's rotation. The term optokinetic reflex encompasses both nystagmus movements of the eyes and the following-behavior of other species.

In 1964 researchers at Yeshiva University were able to apply the optokinetic reflex to the learned-innate debate. The species to be tested, newborn guppies or freshly hatched preying mantids, were exposed to the rotating striped drum and their behavior recorded. To insure that the subjects of the experiment would have no prior opportunity to see, they were all hatched in darkness and

not exposed to light until placed in the device. In every case the animal suspended in the device moved in the direction of the rotating drum (and even reversed direction when the direction of the drum's rotation was reversed), thereby confirming the hypothesis that the optokinetic response is inborn in these species.

In an intriguing variation of their first experiment, the researchers replaced the moving drum with a stationary version made up of a series of vertically arrayed neon tubes. A special rotary switch illuminated the tubes in sequence, so that it looked as though a series of white stripes were rotating about the interior of the drum. This was, of course, not real movement but illusory movement, akin to what we see in the "moving" lights of a theater marquee. Animals exposed to the modified drum responded exactly like those that had been in the real movement condition. Two important conclusions emerged from these experiments: the animals' ability to respond to the rotating drum was innately determined, and no behavioral distinction was made between real and illusory movement.

Do the same principles apply to human beings? Yes. In 1966 much the same experiment was performed, but with human infants as subjects. Some were as young as 10 hours old and none was older than 4½ days. More than two-thirds of those tested on the illusory movement drum showed the visual nystagmus response typical of children and adults. It is highly unlikely that the response could have been acquired in infants so young.

The black-and-white striped pattern included in this book can be used to demonstrate both the optokinetic reflex in a small animal and the corresponding visual nystagmus of a human observer. I suggest that for the first part of the experiment you use an easily obtainable animal, such as a goldfish, although virtually any small animal will do. Construct the striped drum according to the instructions provided, being certain that the stripes are on the inside surface of the drum. Place the assembled drum on a record turntable set to revolve at its lowest speed. You'll find it most convenient to keep the fish in the plastic bag in which it was packed at the pet store; just hold the bag containing the fish in the center of the slowly revolving drum and note the fish's re-

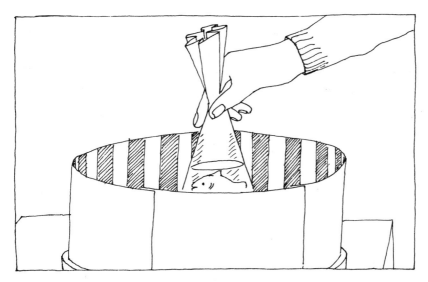

The optokinetic reflex. Assemble the drum so that the stripes are on
its interior. As the drum rotates, suspend a small animal within a clear
container into the drum's interior. Be certain that the container iself
doesn't move (a plastic bag containing a fish is shown here). The animal should move in the same direction as the drum's direction of rotation.

sponse. If it is possible, try rotating the turntable in the opposite
direction in order to reverse the animal's movement.

To demonstrate nystagmus, reassemble the drum so that the
stripes are on the outside surface. Once again the drum will be
rotated on a record turntable. Ask someone to sit a few inches
from the moving turntable and stare at the striped surface. As he
does so, observe the characteristic back-and-forth eye movements
of nystagmus. Any number of modifications to the basic experiment might be tried. For example, you could illuminate the revolving drum intermittently with a flashing light in a darkened
room to produce the discontinuous illumination required for illusory movement. You should expect to see the optokinetic reflex
under these conditions as well.

8. Kinetic Art

A strong resemblance can sometimes be found between the products of visual science and those of visual art. These similarities are understandable since both disciplines strive in their own ways to comprehend the appearances of our world.

The distinction between artist and scientist is a relatively recent historical development. Living in a highly technological culture as we do, it is tempting to assume that science has had a disproportionate influence on art. The reverse is equally plausible, however, with artists supplying the raw material for scientific hypothesis-making. More often, the two groups are merely unaware of the other's existence, and it is my contention that the occasional correspondences between art and visual science, while interesting, have been largely coincidental.

An excellent case in point is to be found in rotating drawings created by both psychologists and artists over the last fifty years or so, all of which exhibit common features. In each of them, a clearly rigid, two-dimensional drawing is made to deform or appear three-dimensional when rotated on a turntable. Strictly speaking, the spirals of Chapter 4 could be included here, since the apparent expansion and contraction of the spiral in motion can be considered complex deformations of this type.

There isn't much point in asking who invented these figures since they have been repeatedly reinvented and rediscovered over the last sixty years. And to complicate matters, new discoverers of the patterns, unaware that others preceded them, or else convinced of the uniqueness of their work, were in the habit of coining new names for the same basic phenomenon.

In psychology, the first to publish a description of the rotating figures was C. L. Musatti, in 1924; but Musatti in turn cited a figure created three years earlier by a colleague named Benussi. The two main types of rotary effects, the apparent deformation of rigid forms in rotary motion and depth produced by rotary motion, were shown respectively by Musatti and Benussi, and the effects sometimes bear their names. It was Musatti who called these effects *stereokinetic*.

In art, a notable collection of rotating disks was published by Marcel Duchamp in 1935, which went through several subsequent publications, some issued as late as 1965. These *rotoreliefs,* as Duchamp named them, were six cardboard disks intended to be viewed while revolving on a record turntable at 33 rpm. All of the disks in this section can be cut out and observed in this manner.

Benussi's figure, while simpler in design, bears a strong likeness to Duchamp's rotoreliefs, but there is no reason to believe that the two were causally related. In both, a series of eccentrically arranged rings seem to pop into relief when rotated. Typically, observers report that the figure looks like a tube extending into the turntable, or else a truncated cone floating above the surface of the turntable. The Duchamp figures incorporate several unexpected and aesthetically pleasing variations on the Benussi effect, which the reader can discover on his own. Some of the effects are ambiguous. For example, the rotating circles will spontaneously change their three-dimensional organization the longer you look at them.

Frederick S. Duncan, who has had training in both psychology and art, has exhibited several variations on Benussi's effect. The acrylic painting "Leaning Tower" is one of Duncan's more persuasive works, which he collectively calls examples of *kinetic art.*

Several factors have been found to influence the appearance of depth in these disks, in addition to those already mentioned. The

Marcel Duchamp's *Rotoreliefs* (1935). Center a pattern on a record turntable set to rotate at 33 revolutions per minute. Although the originals were in color, these black-and-white reproductions will produce a vivid depth effect.

From Marcel Duchamp, *Play-Toy: Six Roto-Reliefs* (#1, #3). Yale University Art Gallery. Gift of Collection Société Anonyme. Reprinted by permission.

ROTORELIEF No.3 — LANTERNE CHINOISE — MODÈLE DÉPOSÉ

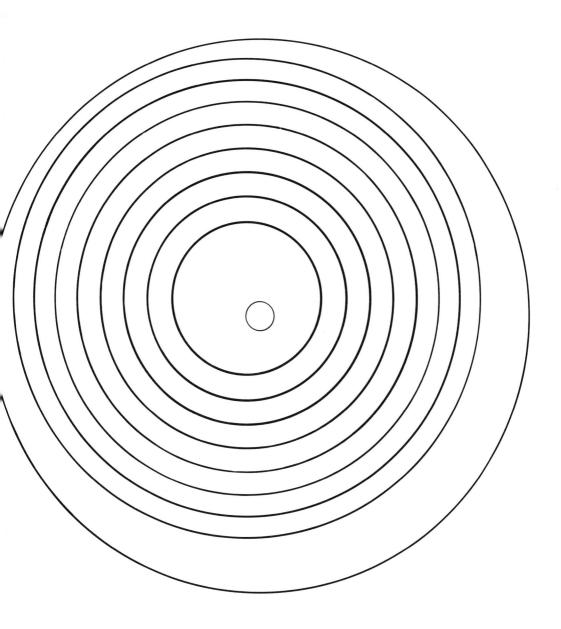

Benussi's figure actually predated Duchamp's and was the first to show apparent depth when rotated on a turntable.

The depth effect is apparent in this adaptation of artist Frederick Duncan's work "Leaning Tower" (1972). It too is intended to be rotated.

Musatti's figure is an ellipse on a disk that seems to deform as the disk rotates, although some observers report having difficulty in obtaining the effect.

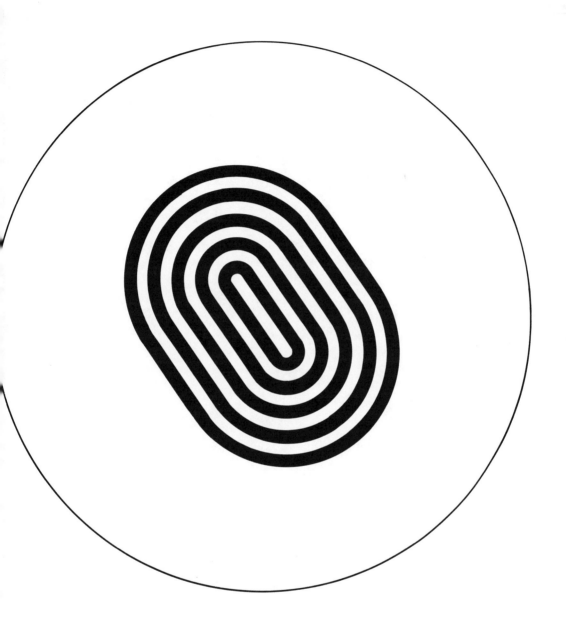

The deformation is much easier to see in this drawing, based on Duncan's 1972 work "Racetrack."

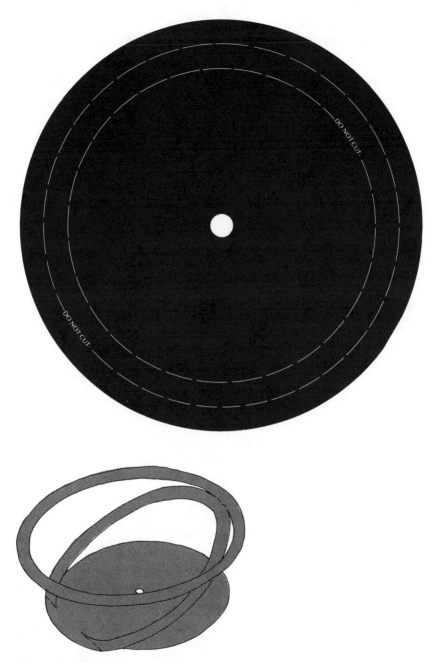

The author noticed this oddity in a store display. Cut out the concentric rings but leave them attached to one another at the points indicated. When the rings are folded as shown and placed on the moving turntable, they will seem to wobble like a spun coin coming to rest.

depth effect can be hastened or intensified by viewing the disk with one eye or by squinting. Squinting may be helpful because the effect is more noticeable when surrounding objects are eliminated. Betty A. Wieland and Roy B. Mefferd, Jr., also found that the appearance of depth in the figures increases with increasing speeds of rotation. You can see this for yourself by looking at the drawings as they revolve on the record turntable at speeds other than 33 rpm.

The deforming effect of rotation, first mentioned by Musatti, is not always as compelling as the depth effect. Musatti's original figure of a solid white ellipse on a black disk may have to be followed for some time before it will seem to wobble about on its axis. In an experiment, Wallach and his colleagues found that specific instruction was critical if most of their subjects were to see deformation of the ellipse in Musatti's disk. On the other hand, a more recent version of the effect created by Frederick Duncan can be seen with no difficulty and without detailed instructions. The picture called "Racetrack" is made up of concentric ellipses that seem to gyrate and shear when centered on a moving turntable.

There is a slightly different rotating deception that I recently saw as part of an advertising display in a shop window. I don't think it has ever been the object of experimental study even though it is probably related to the stereokinetic effect. The rotating figure is made up of two rings, joined at a point on their perimeters and extending vertically in three-dimensional space. A paper version can be made from the pattern I've supplied. When this form is rotated, a motion like that of a spun coin coming to rest is seen, with one halo wobbling about on the perimeter of the other. Because the circles are in a genuine three-dimensional arrangement, it isn't clear whether we are dealing with a Benussi type of effect or if other factors are involved.

9. More Kinetic Depth

The rotating disks of the last section demonstrated the intimate relationship between motion perception and depth perception. In this unit we will look at some other approaches to understanding this relationship and a few of the factors that operate in it.

The *kinetic depth effect* is a concept originated by Hans Wallach and D. N. O'Connell some years ago to describe transformations of moving two-dimensional displays that appear to be three-dimensional to the observer. The effect was originally produced by casting a shadow of an object in motion onto a translucent screen and then asking the viewer to report what he saw. Because observers saw only the shadow of the moving form and not the object directly, whatever depth or solidity was seen on the screen had to have been a function of the changes in the configuration of the shadow, which in turn were induced by the rotational motion of the hidden form.

So that you can better appreciate the effect, build the shadow-caster, being careful to follow the directions exactly. Assemble the screen and the three-dimensional prism; place the prism on a slowly revolving turntable, and observe its shadow on the paper screen. You should have no difficulty identifying a rotating form from what you see on the screen. The ease with which you make the identification may even be misleading, for the important point here is that you did not perceive a prism by viewing the object di-

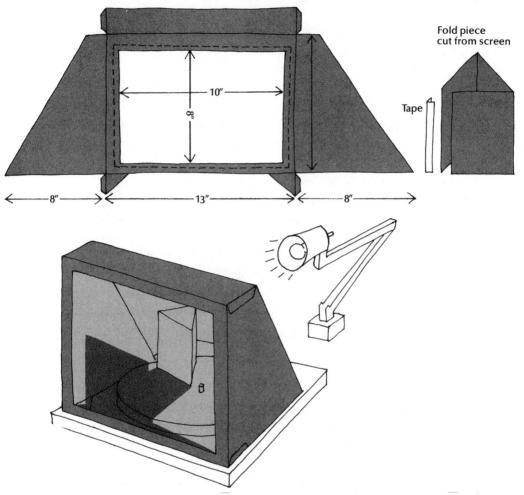

Fold piece
cut from screen

Tape

Build a shadow caster to experiment with the kinetic depth effect. Cut out the frame and tape or glue a sheet of plain paper to it. By folding back the sides, it can be propped up alongside the turntable.

Next assemble the paper prism. Place it on the turntable with a strong light behind (a high intensity lamp works well) so as to cast a shadow of the prism onto the screen. At a given moment the shadow may be rectangular or trapezoidal, a fact you can see by slowly moving the turntable by hand. Yet the constantly deforming shadow seen on the two dimensions of the screen as the turntable revolves at 16 rpm looks like a three-dimensional object in motion.

Experiment with other simple objects, like a bent paper clip or an ordinary teacup.

rectly but by looking at a constantly shifting two-dimensional pattern of light and dark. A significant feature of the demonstration is that the shadow figure will appear to change its direction of rotation as you view it, or it may seem to oscillate back and forth. A variety of shadows cast by revolving objects exhibit this property, and you may want to experiment by projecting the shadows of common objects onto the screen. I've found ambiguous rotation in the shadows of bent paperclips and even a revolving teacup!

The uncertainty of rotational direction is explainable exclusively in terms of what is depicted on the screen. If you take the time to analyze the transformation of the shadow over time, you will see that the same two-dimensional pattern can be produced by an object moving in different ways. Much the same uncertainty of direction of motion had been described by W. J. Sinsteden in 1860. In his account the rotating blades of a windmill, silhouetted against a bright sky, appeared to spontaneously reverse their direction of rotation. A related effect can be seen in a silhouetted Ferris wheel or one whose perimeter is illuminated at night with neon tubes.

While the ambiguity of rotational direction is directly explainable by careful analysis of the shadow pattern, the pervasive depth effects of the patterns are more difficult to understand. Wallach and O'Connell maintained that two conditions must always be met in order for the kinetic depth effect to succeed: the shadow must be displaced across the surface of the screen, and at the same time the shadow must lengthen or shorten. Changes in length alone, they argued, will not produce the effect. Some minor amendments have been added to the two principles since Wallach and O'Connell first stated them in 1953, but they remain the best explanation of the conditions that produce the appearance of solidity in moving, two-dimensional patterns.

What about the stereokinetic disks, like Duchamp's rotoreliefs, that you saw in the last chapter? Are they also an instance of the kinetic depth effect? Wallach, Weisz, and Adams found that the Benussi effect—the eccentrically arranged revolving circles—is attributable to the kinetic depth effect, but they did not find that the distortions of the Musatti ovals are, too. Although both of the earlier effects involve rotation, there is no compelling reason to believe they are expressions of the same perceptual mechanism.

These principles of operation, though not always explicitly acknowledged, are central to the depth we see in all moving, two-dimensional displays. Included here are not only television and motion pictures, but the newer computer animation techniques as well.

One of the most widely reproduced examples of kinetic depth was contrived by Adelbert Ames, Jr., in 1951 to support his views on the central role of past experience in visual perception. Subsequent researchers have generally neglected the theory and concentrated instead on his novel demonstration, the *Ames window*.

The "window" is actually a two-dimensional trapezoidally shaped cutout, made to look like a multi-paned window in perspective. Shaded areas around the panes emphasize the relief or thickness of the window. When the Ames design is slowly rotated about its vertical axis, it will be seen to oscillate back and forth through a path of 180 degrees, instead of its true rotational path of 360 degrees.

As you view the Ames window, remember these hints: The window is meant to be viewed directly, not as a shadow cast on a screen. Although the effect can be obtained with binocular (two-eyed) viewing, monocular vision is more effective. The strength of the illusion varies with distance, the relationship being that it is more powerful when seen at a greater distance. The original pattern was rotated quite slowly, about three to six revolutions per minute; if the slowest speed on your record turntable seems too fast, have someone else rotate the Ames window by hand. In viewing the effect, be patient. It has been my experience that people vary in the time it takes them to experience the effect, although the process will be considerably hastened if you start with monocular viewing. With practice you should be able to see the illusory oscillations of the window almost at will.

To emphasize the incompatibility of the window's real motion with its illusory motion, observe the rotating window as before, but with a solid object passed through the center of the window. A pipe cleaner can be bent and suspended through the central

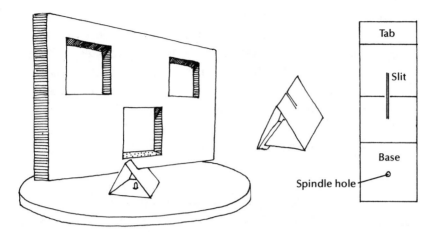

The Ames window. Mount the two window patterns on cardboard, back to back, and attach it to its base as shown in the assembly diagram. The text describes the illusory movement seen in the window as it revolves.

pane, or a length of wire can be poked through the pattern for the same purpose. Since the solid object will continue to move in the circular path while the window appears to oscillate, a perceptual conflict is created that most observers resolve by seeing the window move to and fro while the wire appears to bend and float in the opposite direction.

The observed oscillation of Ames's window is understandable if we assume that the figure is seen not as a trapezoid, but as a rectangular window in perspective. The short end of the window, whether it is physically near or far away from you as it revolves, will always appear distant if interpreted consistently with perspective. Under these circumstances, the true rotational movement of the window will be interpreted as oscillation.

Several variables have been examined for their bearing upon the Ames window. It is accurate to say that the same effects can be found for many rotating objects, not just the trapezoidal window (a point made earlier with the rotating teacup). For example, a wire outline of a trapezoid will also produce the illusion, although Ames pointed out that the less detailed form is less ef-

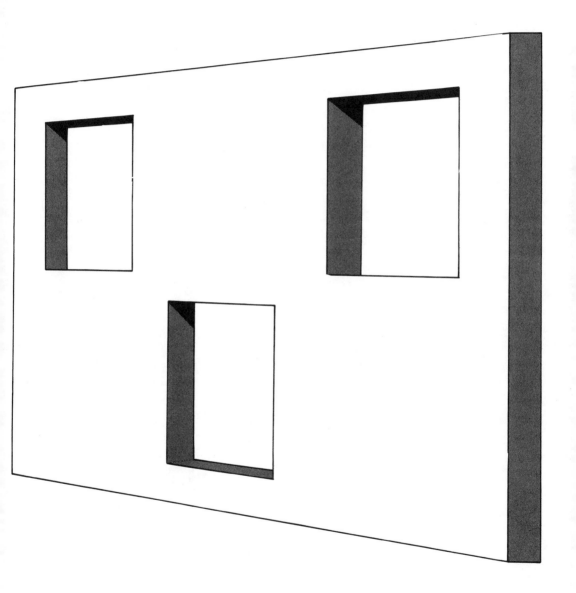

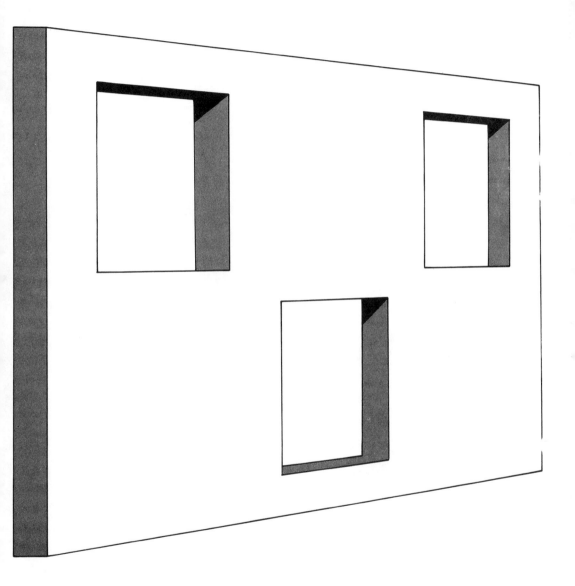

fective. Myron Braunstein and others have studied the effect using computer animation techniques and have found that the length of the vertical edges of the form and the convergence angles of the horizontal lines are critical determinants of the type of motion an observer sees in rotating figures. In general, the more the rotating form resembles a trapezoid, as opposed to a rectangle, the more likely it is that it will be seen to oscillate.

This analysis of the figure is not inconsistent with Ames's own interpretation of the phenomenon. It was Ames's view, however, that the preference for seeing the trapezoidal cutout as a rectangular window in perspective is attributable to our past history of interaction with angular structures. A corresponding theory has been proposed to explain why the geometric illusions are misjudged, a proposal that people reared in angular, "carpentered" environments are more susceptible to the illusions.

Unfortunately, these theories have been resistent to testing. They predict that the oscillation of the Ames window should be more powerful for someone raised in an angular environment than for a person reared in a nonangular world. This specific hypothesis has yet to be tested, but similar ones have produced equivocal results.

Recently, researchers at the University of Minnesota tested infants' reactions to depth stimuli, including a version of the Ames window. Infants have a natural tendency to reach out and touch nearby objects, a reflex useful for measuring whether the child sees an object to be near or far. The behavior was measured in children ranging from 20 to 36 weeks of age for a rectangular cutout and for the Ames window. The rectangular form was situated in such a way that one end or the other was always positioned near the child's hand, while the trapezoidal window was located so that its ends were equidistant to the child's grasp. All of the infants responded to the rectangular cutout by consistently reaching for the near side, but not all reacted to the perspective of the Ames window. The researchers found that reaching for the apparently near side of the Ames window did not develop until somewhere between 22 and 26 weeks of age.

Although the response to perspective did not occur until considerably after birth, this experiment failed to specify whether the reaching response to the Ames window was learned, a function of maturity, or of genetics, or due to a combination of factors. It may well be that Ames was correct in his belief that children must learn the meaning of depth depicted by perspective by interacting with their environment, but it is doubtful that he ever imagined the learning could be accomplished within the relatively brief span of the first six months of life.

10. Leonardo's Window

In the last two chapters, depth perception was considered primarily as a consequence of movement. But since we can readily perceive the distances of objects when no motion occurs, additional factors must be at work. Our ability to see depth is a puzzling aspect of vision, for there is little about the structure of the eye that predicts the uncanny accuracy of human depth perception. The retina is in effect a two-dimensional screen. But since we see the world in three dimensions, there must be something about the two-dimensional retinal image that might well depict the distance of objects in our world. The search for the relationships between the retinal image and our ability to see in the third dimension began in Renaissance times.

The Renaissance dates from the 14th century in Italy, and extended through the 16th century. During this time, science was not a discipline apart from other fields of study. Many subjects that we now consider to be of a scientific nature were taken then to be within the province of philosophy or art, particularly the art of painting. The Renaissance painter's interest in the accurate portrayal of the external world inevitably led him to ponder problems of anatomy, botany, physics, and vision. The difficulties inherent in representing a three-dimensional world in the two dimensions of the canvas held special fascination for painters of the period and forced many of them to consider in detail how depth might be portrayed. Many artists took an approach to the problem guided by the advice of Leon Battista Alberti, who said that a picture should be drafted as if it were a "window through which we look out into a section of the visible world."

Alberti's instruction was taken literally by a number of painters, most notably Leonardo da Vinci, who advised would-be artists:

> Observe at every hundred braccia some objects standing in the landscape, such as trees, houses, men and particular places. Then in front of the first tree have a very steady plate of glass and keep your eye very steady, and then, on this plate of glass, draw a tree, tracing it over the form of that tree. . . . Then, by the same method, represent a second tree, and a third, with the distance of a hundred braccia between each. And these will serve as a standard and guide whenever you work on your own pictures, whereever they may apply, and will enable you to give due distance in those works.

Such a tracing of a scene on glass came to be known as *Leonardo's window*, even though there is little reason to believe that he originated the practice. The actual arrangement of the Leonardo window was depicted in several woodcuts by Albrecht Dürer. You can see that some variation was common in the device; the Leonardo window might well have stimulated the widespread use of drawing aids among artists, a practice that continues today. Although contemporary artists substitute a photographic projection for the display on the glass, the principle is identical.

Before we consider what the Leonardo window has to do with depth perception, try your own hand at making such a drawing. First find a grease pencil or felt-tipped marker that will write on glass (the Sanford brand *Vis-à-vis* marker is a favorite of mine because it won't crawl when used on glass, yet can be wiped off with a damp cloth). You may prefer to draw on a sheet of tracing paper taped to the glass, in which case any writing instrument will do.

Choose a window that affords a varied scene, including trees and buildings. It is extremely important to always keep your eye in the same position as you draw, for otherwise the scene will shift on the glass as you move your head. Elaborate peepholes and

A woodcut from Albrecht Dürer's *Work about the art of measurement* (1525) shows the arrangement of a typical Leonardo window. The peephole was used to maintain the artist's eye in a fixed position.

From Willi Kurth, ed., *The Complete Woodcuts of Albrecht Dürer*. New York: Dover, 1963.

sighting devices for this purpose were fashionable among the artists of Dürer's time, but a simple hole (the smaller the better) punched in a piece of cardboard and taped to the edge of a table will suffice. Remember to always view the scene with the same eye.

Variations were common. For example, a grid fitted to the window allowed what was seen to be transferred to a corresponding configuration on the drawing itself.

From Willi Kurth, ed., *The Complete Woodcuts of Albrecht Dürer*. New York: Dover, 1963.

An example of a Leonardo window you can make. The peephole is punched in a sheet of cardboard fixed to the edge of a table.

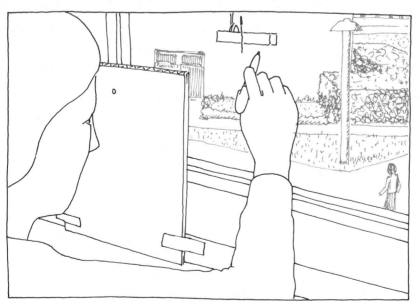

To get a sense of the spirit of the times, I have stressed making the drawing much as it was made by Renaissance artists. In principle, however, the same effect could be achieved by tracing the image produced by a slide projector or overhead projector.

The method you select is less important than the analysis of the picture. As you examine your drawing ask yourself what there is about the picture that suggests the third dimension. What does this have to do with depth perception? Since the retinal image, like the drawing, is a projection of the world onto two dimensions, it must be the case that these features are available to human vision as well. My choice of words here is deliberate, because it is one thing to say that this information is available but quite a

How such a drawing looks, with some of the pictorial depth cues specified.

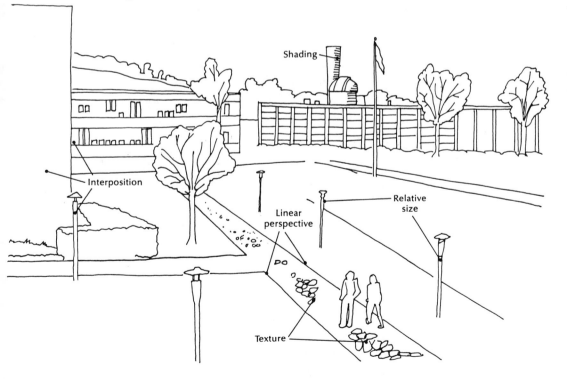

different matter to state categorically that it definitely is employed in vision.

Beginning with Leonardo, commentators have analyzed the so-called *distance cues* present in the plane projection of the Leonardo window and retinal image, all of which are routinely discussed in elementary perception textbooks. See if you can locate the following distance cues in your own drawing:

1. *Linear perspective.* Lines physically parallel to one another in the real world appear to converge upon a common point in the drawing as they recede into the distance. As noted in the last section, linear perspective is the dominant depth cue of the Ames window.
2. *Relative size.* The images of objects become smaller on the drawing as the object becomes more distant in the real world. This subject will be considered in greater detail in Chapter 11.
3. *Gradients of texture.* The psychologist James J. Gibson has pointed out that uniform textures become more compact or compressed in the image as a function of distance.
4. *Interposition.* On the projection, the outline of a nearby object often obscures a portion of more distant objects.
5. *Shading.* Seemingly two-dimensional objects are said to acquire solidity in the presence of their shadows. These objects can also be located in three-dimensional space according to the position of the shadows they cast.

The first three of these distance cues are undoubtedly expressions of a common principle, that the size of an object's image on the retina is proportional to the distance between that object and the observer's eye. Linear perspective, relative size, and gradients of texture all acknowledge that the image on the retina becomes smaller and more condensed with increasing distance. The size of the retinal image plays a key role not only in distance perception, but size perception as well.

A direct measure of retinal image size requires elaborate optical instruments, not generally available to most people. As luck would have it, though, there is an indirect measure that is easily obtained and is perfectly proportional to retinal image size. *Visual*

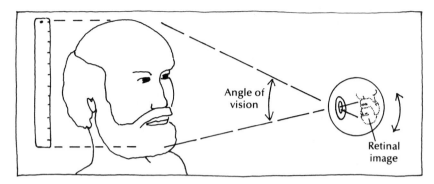

Visual angle is an exact correlate of retinal image size. It is found by extending two lines from the extremities of an object through the center of the eye's lens. Size terminology can be confusing. The physical size of an object is determined with a measuring instrument, as is retinal image size, usually by calculating visual angle. Apparent or perceived size, on the other hand, is judgmental.

angle, as it is known, is obtained by drawing two lines from the extremities of an object and passing them through the center of the ocular lens. It's easy to see that as the angle formed at the intersection of the lines increases, the size of the image also increases. Conversely, as the visual angle diminishes, so too does the size of the retinal image. It has become common practice to substitute visual angle for retinal image size.

Because the indicators of depth illustrated in Leonardo's window are present in drawings, paintings, and photographs, they have been called the *pictorial depth cues,* or the *static depth cues* since motion is not necessary for their operation. Not everyone agrees upon their role in depth perception, and some theorists even argue that describing depth perception in terms of isolated cues is pointless since they may not operate in nature as they do in paintings.

Even if we assume that human beings employ pictorial depth cues in judging distance, we still do not fully understand how they are established. Most evidence now indicates that they are learned. Nor do we understand the subtle interactions that must occur between the pictorial cues and other sources of information about depth.

11. The Size Problem

By now you have undoubtedly sensed that perceptual researchers enjoy studying problems that most other people would prefer to ignore. I am sure very few people lose any sleep over what it is that makes something a certain color or shape, or why it appears to be a certain distance away. I also assume that few of us ever wonder why an object seems to be a given size: Why is it that a playing card looks as if it is 4 inches high while the person holding it looks as if he is 6 feet high? This question of how it is that we can correctly identify the sizes of things is the broadest possible statement of the size problem.

When I've introduced this subject to my students, many have found it convenient to explain size perception on the grounds that things must have a certain size because we know how big they are. But to say that we "know" how big things are isn't much of an explanation unless the definition of knowing can be pinned down. Presumably, knowing refers to the fact that we have had past experience with objects. That is, we have learned the dimensions of objects by interacting with them. How? What is the nature of this educational process? How important is it to the perception of size? If learning were the sole basis for size perception, then a person would never be able to specify the size of an unfamiliar object the first time he saw it. Since this is patently not how things work, we can assume, on logical grounds alone, that more than learning is involved in size perception.

This examination of how it is that we judge size as precisely as we do is more than just an inquiry into the innate-versus-learned issue. It is also a search for those variables within the organism and within the stimulus that make size perception possible.

The two most crucial elements in our ability to perceive size are: 1. the size of the image that an object creates on the retina (the retinal image size or visual angle) and 2. the distance of that object from the eye. It is important to understand why retinal image size alone is insufficient to account for size perception.

SIZE CONSTANCY

Suppose for a moment that the only information available to the visual system was the image spread across the curved surface of the retina, and the object's distance could not be judged. What then would happen to that image as its owner moved toward or away from the object that gave rise to it? When the object was close to the eye, its image would be larger than when the object was more distant. If size perception were determined *only* by the size of the retinal image, then the world would be a most chaotic place, a plastic universe in which the sizes of objects would shrink and grow as our own distance from those objects changed. The fact that size remains stable in perception, despite fluctuations in the size of the retinal image as we move about, is testimony to the existence of *size constancy*, and this constancy of perceived size exists because our visual systems also take distance into account. Mathematically, it is a simple business to predict the apparent size of an object by incorporating these two factors, retinal image size and distance, into an equation. We must assume that a comparable process occurs within our brains.

Let me pause here to untangle the inevitable complications in terminology that arise in any consideration of size perception. We can always assume that an object has a *physical size*, which can be quantified with the appropriate measuring instrument, such as a ruler. The size judgments made by human beings (and other species) are almost always in agreement with the physical size of things, but it must be stressed that the two are not always in per-

fect agreement. Significant discrepancies may occur. For this reason it is necessary to distinguish between physical size and *judged
size*, which is also called *apparent* or *perceived size*. Retinal image
size, as noted above, can also be measured as visual angle, which
of course is not the same as the physical size of the object itself,
nor is it always an exact correlate of perceived size.

One way to appreciate the interaction between distance perception and visual angle is to observe objects when distance cues
have been minimized. As a practical matter, it isn't easy to oblit-
erate all evidence of distance. Usually this is done by having
someone judge the size of a luminous object within a completely
darkened laboratory, while looking through a tiny aperture, called
an *artificial pupil*. Since our demonstrations need not be so rigor-

When distance cues are minimized, both size and distance perception
may be affected. Observe the white rectangle on a black field through
an artificial pupil, the latter made by poking a pin through an index
card. The form should be held in such a way that only the center of the
page will be seen while the edges of the page remain an indistinct blur.
See the text for the role of instructions in this demonstration.

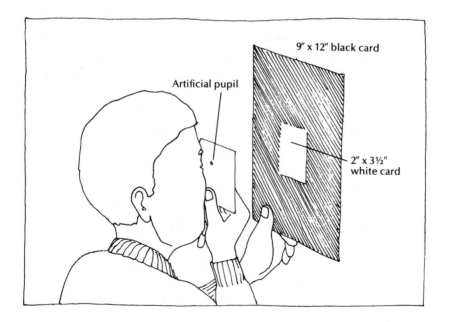

ous, I've devised a simpler technique. By viewing an object through an artificial pupil (a pinhole poked in a 3 × 5 index card) while the object's surroundings are screened from view, you should have some appreciation of how things look in the absence of distance cues. The size of an object may well appear indefinite under these special *reduction conditions*—and, of course, so will its distance. As you examine the white rectangle while looking through the artificial pupil, it may seem to be quite close to you, far away, or at some intermediate distance. Albert H. Hastorf has noted that, as with other ambiguous stimuli, the appearance of the white rectangle is readily influenced by instructions. Look again at the rectangle but this time imagine that it is a sheet of typing paper. How far away does it seem to be? Then imagine that it is a small calling card and ask yourself how its apparent distance has been changed.

In 1881, E. Emmert showed in convincing fashion, just how size perception is affected by the relationship between distance and visual angle. *Emmert's Law* describes the dependence of the perceived size of an afterimage upon the distance of the surface on which it is viewed. *Afterimages*, like the spots we see after looking at the flare of a flashbulb, are assumed to occur because the retinal receptors become fatigued for a time by an intense or enduring light source, even after that source ceases.

Emmert's law noted that an afterimage viewed against a distant surface appears larger than the same afterimage seen against a nearby surface. This principle is more understandable if you see a practical demonstration of it. One way to do so is to form an afterimage on a distant surface, such as a wall, and then on a sheet of paper held at reading distance. The afterimage can be formed by staring for about a half-minute at a shaded lamp. A mask with a distinct shape placed over the lamp, as illustrated, may make the demonstration more vivid, but is not essential. Make note of how the size of the afterimage seems to change depending on the distance at which it appears to be. As you experiment with afterimages, keep in mind that it is not necessary to stare at an intense light source in order to form an afterimage.

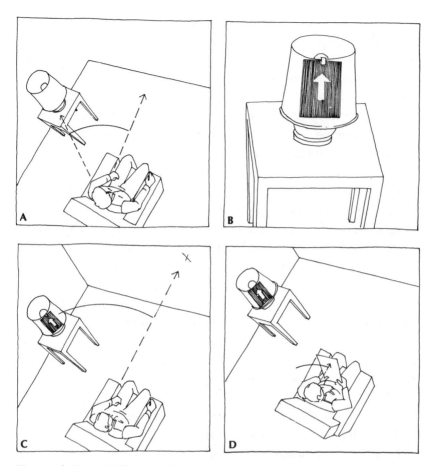

Emmert's Law is illustrated by viewing afterimages on surfaces at different distances:

A. Stare at a shaded lamp for 30 seconds or so and then shift your gaze, but continue staring; a dim afterimage will appear for a few seconds and then fade away.

B. The afterimage will be more distinct if you first cover the lamp with a paper mask from which a shape has been cut.

C. Form an afterimage on a relatively distant surface, such as a blank wall, taking note of the apparent size of the fleeting afterimage.

D. Repeat the procedure, but this time form the afterimage on a sheet of blank paper held at reading distance. How has the apparent size of the afterimage changed?

UNDER NO CIRCUMSTANCES SHOULD YOU ATTEMPT TO STARE AT THE
SUN OR AT A HIGH INTENSITY LAMP WHEN EXPERIMENTING WITH
AFTERIMAGES!

It is inherent to Emmert's law that whenever the size of the
retinal image remains constant, the perceived size of the object
will seem *larger* as the distance to the object increases. In everyday life, of course, the visual angle of an object does not remain
constant with changes in distance, i.e., as the object recedes into
the distance, its visual angle actually diminishes. So this application of Emmert's law only holds for those occasional circumstances when the visual angle remains constant but distance
changes.

Carrying this line of reasoning a bit further, it is also the case
that in the real world, when two objects subtend equal visual angles, but are at different distances, then the more distant object
not only *looks* larger, but it *is* physically larger than the nearby
object. An example of this might occur when an adult is seen
standing a few feet farther away from us than a child. Both cause
the same-sized images to form on our retinas (the objects subtend
equal visual angles), but the adult looks larger than the child, and
the adult *is* physically larger than the child.

I've emphasized these relationships in order to clarify some of
the confusion surrounding textbook demonstrations of Emmert's
law, particularly the classic *corridor illusion*. The corridor illusion
typically takes the form of a drawing of a stylized hallway in
which are located two solid geometric forms (cones or cylinders),
although some versions substitute cartoon figures. The observer is
asked which figure appears larger, and invariably it is the more
"distant" figure that looks bigger. The figure is claimed to be an
illusion because when the reader measures the two cones with
a ruler, they are found to be the same size. The corridor illusion
is not so much an illusion as it is a misleading problem. Let's
see why.

An illusion, as the word is used in perception, occurs when the
judgment of a stimulus does not correspond to a physical measurement of that stimulus (see also Chap. 19). In the corridor
"illusion," it is not clear whether we are to see the stimulus as two-
dimensional or three-dimensional. In any 3-D interior, the only

Perceived distance may play a role in the corridor illusion. For example, compare the apparent size of the cones with and without the per-
spective-rich background present. In principle, however, the illustration is no different from any plane projection of a three-dimensional
world, and in this sense is no illusion.

way that the distant figure could look larger than the closer figure would be if it were actually physically larger (i.e., the judgment of the stimulus would correspond to its measurement). The fact that the cones subtend equal visual angles on the observer's retina is coincidental.

The "illusion," therefore, depends on the observer being misled into seeing the stimulus as 3-D and making a (correct) judgment

Familiar size implies that learning plays a role in size perception. When the three playing cards are viewed through the artificial pupil, most observers report them to appear at different distances; this may have to do with our familiarity with playing cards or because we make comparisons among these conceptually identical objects, the latter being an instance of a relative size judgment.

on that basis. The trick is that the observation is then "disproved" in 2-D. This arbitrary requirement disqualifies the problem as an illusion.

To make matters worse, nobody has ever bothered to make absolutely certain that it is the pictorial depth cue of perspective that creates the effect of depth in the hallway and, by extension, the size of the cones. It is also possible that the density of the textures surrounding the forms alters their apparent size. As things stand, the corridor illusion may possibly be an instance of Emmert's law at work, but it is no more remarkable than a photograph of a child with his parents in the background.

LEARNED SIZE

To return to the problem with which this section began, questions still remain about the role of learned or familiar size. As it happens, not only does the distance of an object tell us something about its size, but the size of an object tells us something about its distance. The reciprocity of size and distance has been of central importance to several researchers, including Ames, who stressed the role of past experience in visual perception.

If you think back to the example of the Leonardo window, you will recall that the *relative size* of objects was one of the pictorial distance cues. Relative size is subtly different from familiar size (though the two may be related), implying as it does that by making simultaneous size comparisons among objects, we can infer something about distance, even with no previous knowledge of the object.

To better understand how this principle operates, examine the playing card illustration with the artificial pupil mentioned earlier. Even though other depth cues have been reduced, the cards should nonetheless appear to be at different distances. This demonstration, a simplified version of one created by Ames, may make use of the fact that we have had past experience with playing cards (familiar size) or that we make comparisons among the similar playing card stimuli (relative size). It is also conceivable that both variables are at work.

12. Two Eyes Are Better than One

In the year 1838, Sir Charles Wheatstone, a brilliant and prolific scientist-inventor, presented a paper to the Royal Society in which he described the workings of a previously unknown feature of vision. It was his discovery that would firmly establish a link between depth perception and physiology.

Basically, Wheatstone had speculated upon binocular vision: that there is a reason why human beings, like many other creatures, have two eyes rather than some other arrangement. Two eyes are better than one because they increase the size of the visual field and allow us to "take in" more of our surroundings than would be possible with a single eye. For human vision, under ideal viewing conditions, we can image fully 208 degrees of our surroundings because of the enormous refractive power of the corneas. This means that we can actually see slightly behind ourselves. Of course only a small part of that visual field will be seen with clarity—the one degree of retinal extent that is encompassed by the fovea.

FOVEAL VISION AND BINOCULAR VIEWING

Wheatstone was greatly interested in the fact that the images formed by the two eyes are slightly different from one another (*retinal disparity*), a consequence of the fact that each eye is looking at the object from a different position.

You will recall from Chapter 3 that visual acuity, the ability to discern fine details, is the province of cone cells, most highly concentrated at the center of the retina in the region of the fovea. When objects are at relatively great distances from an observer, they are imaged on the fovea when the axes of the two eyes are parallel to each other. As the object approaches the observer, the eyes turn inward in order to maintain it in foveal vision. The change in the *convergence angle* of the eyes is automatic and we are rarely aware of it unless it is specifically called to our attention. The change in convergence that coincides with the approach of an object can be appreciated by watching what happens to someone's eyes as he stares at an approaching pencil.

At the same time that the eyes converge (so that the image of an object will be centered about the fovea), a second and coordinated change occurs. Since the lens of the eye cannot move in and out the way a camera lens does, the sharp focusing of the image necessary for clear vision must be accomplished by other means, and a unique mechanism of the eye accomplishes this focusing of the image by changing the curvature of the lens, a process known as *accommodation*. The change in the shape of the lens, for which there is no counterpart in optical instruments, is coordinated with convergence, and because these functions are so completely integrated, they can be considered part of the same mechanism. Accommodation and convergence are changes that the eyes undergo that may well be physiological cues to depth, i.e., the changes in physiology may contribute to depth perception. This is also true of retinal disparity, particularly at the level of the brain.

Although others had been vaguely aware of retinal disparity, it was not until the appearance of Wheatstone's 1838 paper that its true significance was realized. You can examine the disparity between your own retinal images by holding any small, angular object at reading distance and blinking alternatively so as to observe it first with one eye and then the other. With regular alternation, the object will appear to jump, a case of illusory movement between the slightly differing images.

Under ordinary circumstances we are unaware of the discrepancy between the retinal images. Unless the two images are extremely divergent, they tend to fuse into a single impression of the world, and it is the hidden occurrence of fusion that may have delayed the investigation of retinal disparity until Wheatstone's time.

STEREOPSIS

Wheatstone devised an experiment by which he could systematically explore the workings of disparity. He drew two views of an uncomplicated geometric object, each of which was in effect a Leonardo window for the left and right eyes. Then, when the left eye's image was presented to the left eye, and the right eye's image to the right eye, the fused image appeared in depth! By combining disparate retinal images, depth perception accomplished what has come to be called *stereopsis*. Sir Charles wrote with restrained enthusiasm of the fundamental discovery that retinal disparity was an important cue to depth:

> the projection of two obviously dissimilar pictures on the two retinae when a single object is viewed, while the optic axes converge, must therefore be regarded as a new fact in the theory of vision.

There is a substantial problem in re-creating Wheatstone's demonstration. If the two pictures (called *stereo half-fields*) are printed side-by-side and held at fairly close range, both pictures will form images on both retinas. The trick is to find a means of segregating the two half-fields, each to its corresponding eye.

One technique that requires no equipment at all is called *free stereoscopy*. The observer must look at the left-hand picture with only the left eye, and the right-hand picture with only the right eye. This is done by relaxing the normal convergence tendency he has in looking at nearby objects, usually by imagining that he is staring straight ahead at a distant wall while looking at the paired pictures. Some stereo pairs, or *stereograms*, are easier to see using free stereoscopy than others. Elliot's *ocular views* of the last century were intended specifically to be viewed in this manner. The white spot partially obscured by the cross acts as a focal point for

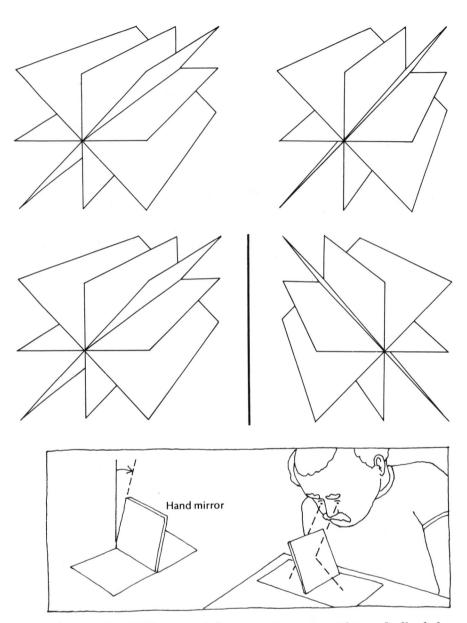

An example of Wheatstone's drawn stereograms. This and all of the other stereograms are presented in two forms, one for viewing with a conventional stereoscope, and the other for viewing with a small mirror. If the second technique is selected, hold the mirror along the line separating the two pictures (or half-fields) and stare at the right picture directly with the right eye while looking at the reflected left image with the left eye. By tilting the mirror slightly, the two images can be made to fuse.

Eliot's ocular views were intended for free stereoscopy. Simply hold the stereogram about ten inches from your eyes with the pictures centered. As you look at the pictures, imagine that you are staring at a distant wall and the two images of the moon should slowly merge until the two half-fields fuse.

the reader to gauge changes in convergence. The method for viewing Elliot's pictures is explained in the figure and with considerable practice may be applied to any stereogram.

The difficulty with free stereoscopy, other than learning how to do it, is that accommodation (focusing) is altered at the same time that the convergence angle changes. Although the half-fields of the stereogram are fairly close to the observer, convergence must be modified as if looking at a distant object, and accommodation will change accordingly. Even though accommodation can be made to follow convergence, the combined image may be uncomfortable to maintain. While squinting may help a little to clarify the pictures, free stereoscopy remains tiresome, even to an experienced viewer.

Wheatstone was among the first to recognize that to view the stereoscopic stimuli with ease, a viewing device would be required, and he promptly invented a mirrored apparatus that he called the *stereoscope*. While this was a definite improvement over free stereoscopy, Wheatstone's mirrored stereoscope was cumbersome and effective only for viewing relatively large stereo-

grams. The mirrored stereoscope was followed by the invention
of a more compact version employing lens segments, the effect of
which was to render the two pictures at optical infinity while
maintaining correct accommodation. It is the lens type of stereoscope that remains in use today, including the popular Viewmaster system with its rotating cards of stereoscopic transparencies.

A photographic stereogram from the author's collection. The stereogram flourished as an amusement at the turn-of-the-century. This example was titled "Among the prize chrysanthemums" and was published
in 1898 by Strohmeyer and Wyman of New York.

For anyone writing about stereopsis—including me—it is always a problem to come up with a cheap and effective stereoscope for the reader to use. As a compromise I have included two versions of each stereogram, one intended for use in a lens stereoscope, and the other for a simpler technique that requires only a small hand mirror. The mirror method will at least allow you to experience stereopsis, but your viewing will be limited to the specially prepared stereograms in this book. For your convenience, I have included at the end of this chapter a list of sources where a stereoscope might be found or purchased.

The invention of the stereoscope coincided with the invention of photography, and the two were combined so readily that drawn stereograms became virtually unknown by the 1860s. So enthusiastically were stereograms regarded, that it was common practice for photographers of the mid-nineteenth century, particularly for those who photographed exotic locales, to take a stereographic view of any scene taken with an ordinary camera. Stereoscopic photography was a booming industry until the advent of the cheaper box camera photograph made it possible for any amateur to take his own pictures. By the early part of the present century, the international mania for stereograms had begun an irreversible decline, such that today stereograms are more of an esoteric collector's item than a legitimate form of entertainment and basis for scientific investigation.

In principle the photographic stereogram is no different from Wheatstone's drawn stereograms. Stereographic cameras are functionally two cameras combined into a single unit, having two lenses separated by the same distance that separates the human eyes, with shutter mechanisms designed to operate in unison. The pair of pictures produced by the camera may be made into prints, as was the practice in the nineteenth century, color transparencies (typical of the Viewmaster), or adapted to any of several commercial processes.

A camera specifically designed to make stereograms is not essential for you to make your own stereograms, although a stereoscope is required to view them. Two techniques are illustrated by which you can make your own stereograms: the crossover tech-

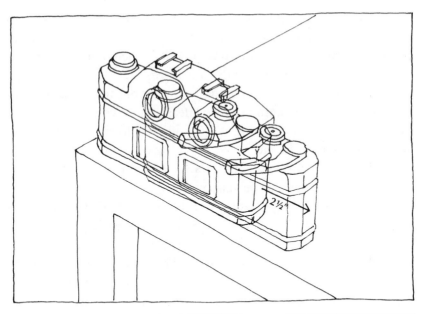

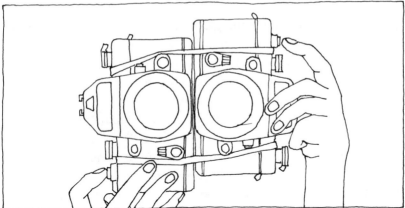

Photographic stereograms for use with a stereoscope are easy to make with any conventional camera. Two methods are illustrated:

A. In the crossover technique, a single camera is used. First one exposure is made, and then the camera is shifted laterally by a distance of at least 2.5 inches. The resultant pictures can be viewed as the left and right half-fields of a stereogram.

B. If two cameras are available, they can be temporarily strapped together. Hold the cameras vertically and trip the shutters simultaneously. In both techniques it is important to keep careful records of which picture was taken by the left camera and which by the right camera so that they can be viewed properly.

nique, and a method in which two identical cameras are attached to each other. In the crossover method, any camera can be used: First one picture is taken and then the camera is shifted laterally 2.5 inches, where the second picture is exposed. This method has the advantage of requiring the minimum of equipment, but alignment is critical and it is useless for taking subject-matter in motion. One way to check the alignment of the camera is to mark off the displacement distance on the edge of a sturdy table and then align the back of the camera with the edge of the table for each exposure. Still simpler is the technique that uses two cameras of identical format (i.e., using the same size film). The cameras are strapped together bottom to bottom with heavy rubber bands, which, for most cameras, will automatically make the interlens distance approximate the 2.5 inch interocular distance. For either method a Polaroid or other instant camera is perfectly acceptable for making stereograms and has inherent advantages for the experimenter.

A simple experiment can be conducted by making several stereograms of the same subject while varying the distance between the lenses of the two cameras, or the extent of the shift between frames if the crossover technique is used. The resultant photographic pairs correspond to retinal images that would have been obtained with comparable interocular distances. The effect of increasing the interlens distance is to increase the disparity between related points in the two pictures, and thus to increase the apparent depth seen in the stereogram. The ability to amplify perceived depth by increasing disparity has been used to advantage in commercial and military applications of stereoscopy, such as the interpretation of stereoscopic aerial photographs.

The elemental nature of stereopsis as a basis for depth perception has been studied extensively by Bela Julesz of the Bell Telephone Laboratories, using novel and technologically sophisticated stereograms of his own contrivance. The stereograms were generated by a computer program in the following way: First the left half-field of the stereogram was generated by producing a matrix of black and white squares, in equal numbers but in a random distribution. The right half-field was identical to the left pattern but the computer shifted a central area relative to the left half-

A Julesz random-dot stereogram. Depth perceived in this stereogram
is attributable solely to stereopsis, since no other depth cues are present.

From Bela Julesz, *Foundations of Cyclopean Perception*. Copyright 1971, Bell
Telephone Laboratories. Reprinted by permission.

field, creating a region of relative disparity between the two patterns. The patterns seen individually appear to be flat textures of
black-and-white, but when placed in the stereoscope, the shifted
region is seen in distinct relief. Julesz's computer-generated, random-dot stereograms show that depth perception can be mediated
by retinal disparity alone, since nothing else about the patterns,
such as the presence of pictorial depth cues, exists in the patterns
to suggest that one area is in relief.

This technique has also been used to test whether certain visual
phenomena are peripheral in nature (attributable to the workings

of the eyes) or central (attributable to the central nervous system). For example, some theories of the geometric illusions attribute these odd misjudgments to characteristics of the viewer's eyes—to factors such as eye movements or contrast effects on the retina. By making a random-dot stereogram of an illusion which can be fully integrated only with binocular viewing, the suitability of the peripheral explanation can be evaluated. Since the optic

This random-dot stereogram depicts the Poggendorff illusion (described in detail in Chapter 19). Because the illusion persists when the half-fields are combined, it has been suggested that the illusion is primarily neural in origin, rather than being a function of the eye.

From Bela Julesz, *Foundations of Cyclopean Perception.* Copyright 1971, Bell Telephone Laboratories. Reprinted by permission.

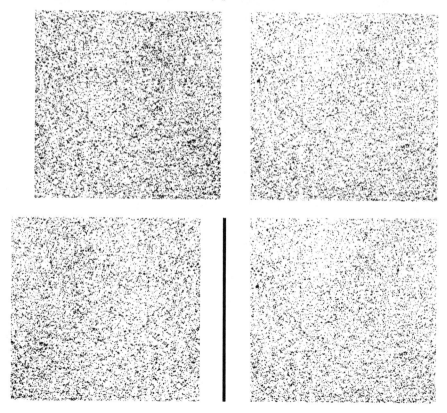

nerves are independent of one another until they reach the brain, an effect perceivable in the random-dot stereogram format must have a central level explanation rather than a peripheral one. In the example shown here, the well-known Poggendorff illusion has been transformed into a random-dot stereogram, one which can only be integrated at the neural level. Since the illusion persists when the pattern is viewed stereoscopically, it can be argued that the Poggendorff illusion is most likely an effect of the brain rather than the eyes.

A stereogram similar in principle to Julesz's but far simpler to make is the typewriter stereogram originated by Lloyd Kaufman, an example of which is also shown. The shift between the two half-fields can be seen by carefully examining the letters of the matrix. You should be able to detect a central square of letters that has been shifted one column laterally. When the complete stereogram is viewed in the stereoscope, the central area will be seen to float above its background. By interchanging the left and right half-fields, the depth effect will reverse—the area that for-merly was seen above its background will now be seen behind it. The reversal of perceived depth when the half-fields are inter-changed is predicted by the geometry of the situation, a topic too esoteric for present consideration, although the basic phenomenon can be used to address several theoretical questions.

One question that has intrigued me concerns what happens in a photographic stereogram when the two half-fields are inter-changed. Do faces suddenly look like indentations and distant buildings stand out from the surface of the photograph? In most photographic stereograms this never happens, as you can see for yourself by interchanging the half-fields of the stereograms pro-vided in this book. It seems to be the case that a photograph rich in pictorial depth cues maintains depth among objects pretty much the same way whether the stereoscopic cue works for or against the pictorial information. Or, to put it somewhat differ-ently, when the stereoscopic cue is in conflict with many pictorial depth cues, stereopsis loses. The depth perception afforded by stereopsis may be a kind of fine tuning of the system, only one of

```
bmcbdjksoaiufhrgdvxkjgheidapmzxytfhwdgcj        bmcbdjksoaiufhrgdvxkjgheidapmzxytfhwdgcj
shyfowmzkleifoayfhslkjwbcywkasnbjhfmkord        shyfowmzkleifoayfhslkjwbcywkasnbjhfmkord
krfbscuhrcspktbqjpfvuthkdcrilmtbnuhfvgdx        krfbscuhrcspktbqjpfvuthkdcrilmtbnuhfvgdx
patxkrbiejlsximudkybfaxkltbdugxlknydjlcx        patxkrbiejlsximudkybfaxkltbdugxlknydjlcx
psfrdcjlgsbmydchksvyfcjkoutrdvgmsckhbkfm        psfrdcjlgsbmydchksvyfcjkoutrdvgmsckhbkfm
utfvnjfxiyjdmeskmrjxugslmgbdxkhrvugsvkgh        utfvnjfxiyjdmeskmrjxugslmgbdxkhrvugsvkgh
ufbsjlbdfvgwruojgdxcvlmhsbcurjxklyhbrhjv        ufbsjlbdfvgruojgdxcvlmhsbcuyrjxklyhbrhjv
lugdxjnrdsuolmrcgyxsjhipfsnvfwyerxiplkmh        lugdxjnrdsulmrcgyxsjhipfsnvwfwyerxiplkmh
gsunxqphedcuijmsagckpnyvxrfgnudskubfxhjn        gsunxqphedcijmsagckpnyvxrfgenudzkubfxhjn
lumvhdzrtfcsynmigkncezyhikmrtcghjfbexdit        lumvhdzrtfcynmigkncezyhikmrptcghjfbexdit
lhcrsxkybjrdbgjurdxuhvkgrsjncuokldbvrxyj        lhcrsxkybjrbgjurdxuhvkgrsjnxcuokldbvrxyj
lmtvhdubfhujbfxkunhdryjkibdekyhvfjybdcxg        lmtvhdubfhubfxkunhdryjkibdenkyhvfjybdcxg
lmhecshatfbdeesupjkinmbghydxesxwsfdghnvrf       lmhecshatfbdeesupjkinmbghydxesxwzfdghnvrf
jdhgbcxrdsjhbdxukrdhncsiklrvshbfrchudzlm        jdhgbcxrdsjhbdxukrdhncsiklrvshbfrchudzlm
xlfutgrslnuvcbkfhtdxokmrhbdgcrsklmycidwj        xlfutgrslnuvcbkfhtdxokmrhbdgcrsklmycidwj
slkguyvvxiosjhvfegxjmwhrkslnbrfyshbvtdkg        slkguyvvxiosjhvfegxjmwhrkslnbrfyshbvtdkg
mxhdryklsbcjrsjbtwhkvschvurgsouncgbrdsx         mxhdryklsbcjrsjbtwhkvschvurgsouncgbrdsx
lrihsbcyrjyplvstgezgdnklnmtdvfxjspkrwnfy        lrihsbcyrjyplvstgezgdnklnmtdvfxjspkrwnfy
```

```
(stereogram half-field with rotated/scrambled       bmcbdjksoaiufhrgdvxkjgheidapmzxytfhwdgcj
 typewritten letters)                                shyfowmzkleifoayfhslkjwbcywkasnbjhfmkord
                                                     krfbscuhrcspktbqjpfvuthkdcrilmtbnuhfvgdx
                                                     patxkrbiejlsximudkybfaxkltbdugxlknydjlcx
                                                     psfrdcjlgsbmydchksvyfcjkoutrdvgmsckhbkfm
                                                     utfvnjfxiyjdmeskmrjxugslmgbdxkhrvugsvkgh
                                                     ufbsjlbdfvgruojgdxcvlmhsbcuyrjxklyhbrhjv
                                                     lugdxjnrdsulmrcgyxsjhipfsnvwfwyerxiplkmh
                                                     gsunxqphedcijmsagckpnyvxrfgenudzkubfxhjn
                                                     lumvhdzrtfcynmigkncezyhikmrptcghjfbexdit
                                                     lhcrsxkybjrbgjurdxuhvkgrsjnxcuokldbvrxyj
                                                     lmtvhdubfhubfxkunhdryjkibdenkyhvfjybdcxg
                                                     lmhecshatfbdeesupjkinmbghydxesxwzfdghnvrf
                                                     jdhgbcxrdsjhbdxukrdhncsiklrvshbfrchudzlm
                                                     xlfutgrslnuvcbkfhtdxokmrhbdgcrsklmycidwj
                                                     slkguyvvxiosjhvfegxjmwhrkslnbrfyshbvtdkg
                                                     mxhdryklsbcjrsjbtwhkvschvurgsouncgbrdsx
                                                     lrihsbcyrjyplvstgezgdnklnmtdvfxjspkrwnfy
```

The typewritten stereogram is conceptually similar to Julesz's stereogram. Careful study will show that one portion of letters in one of the half-fields has been shifted in relation to the pattern of letters in the other half-field.

From Lloyd Kaufman, *Perception: The World Transformed.* Copyright © 1979 by Oxford University Press, Inc. Reprinted by permission.

several ways to judge the three-dimensional properties of objects at relatively close distances from the observer.

WHERE TO FIND A STEREOSCOPE

If you plan to investigate stereopsis more thoroughly you will need a stereoscope of reasonable quality. At one time stereoscopes were a frequent commodity of antique and second-hand stores, but the increased demand for old stereoscopes has put them beyond the means of most of us. If your library or bookstore has a copy of *Sight and Mind* by Lloyd Kaufman (Oxford University Press, 1974), you will find a stereoscope included with the book along with some very interesting stereograms.

Stereoscopes are still manufactured, primarily for the purpose of aerial map interpretation. Most colleges and universities with geography or forestry departments have stereoscopes on hand, either to borrow or purchase. At the college where I teach, a folding stereoscope is sold in the bookstore for less than two dollars. You could also write directly to the manufacturer for one: Hubbard Scientific Company, P.O. Box 105, Northbrook, Illinois 60062. A much more costly version is sold by The Edmund Scientific Company, Edscorp Building, Barrington, New Jersey 08007. Edmund also sells the optical components for building your own stereoscope.

13. Pulfrich's Amazing Pendulum

Few demonstrations in visual perception can match the *Pulfrich pendulum* for vividness and simplicity. The effect, named for its discoverer, the German physicist Carl Pulfrich, borders on the incredible, even for an observer who has some understanding of the phenomenon. The apparatus needed to demonstrate the effect is minimal: a length of string or thread, a small weight, and an optical filter. A metal housekey can be used as the weight, and almost any colored, transparent material will serve as a filter; a glass from a pair of sunglasses, or a patch of colored glass or cellophane will usually suffice. The color of the filter is unimportant.

To see Pulfrich's pendulum in action, attach the key to a length of string so as to make a pendulum. Then hold the filter over one eye, leaving the other eye uncovered, and swing the pendulum back and forth in front of you at arm's length, observing its path as you do so. If the pendulum were viewed from above, its true path would describe a straight line as it moved back and forth, but it won't look that way with the filter covering one eye. Instead, the pendulum will seem to travel a distinct elliptical orbit, apparently swinging toward and away from you. When the filter is switched to the other eye, the apparent direction of rotation will reverse. It is the peculiar discrepancy between the true path of the pendulum and its perceived path to someone viewing it with one eye filtered that Pulfrich first described.

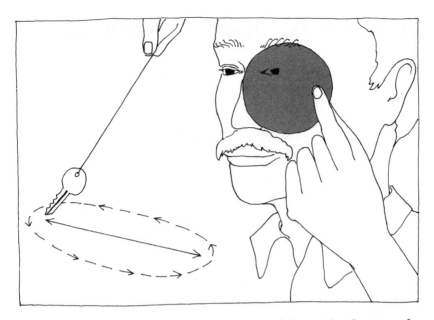

The general arrangement of the Pulfrich pendulum. The drawing also shows the true path of the pendulum bob (solid line) and the perceived path (broken line).

To best see the effect, remember the following points: There is a tendency for some observers to squint while viewing the pendulum, perhaps because the eyes are treated differentially. Both eyes must remain open throughout. There is also an inadvertent tendency to make the pendulum acutally move in an elliptical orbit, which must be avoided. It might be better to have two people work the demonstration, one to carefully swing the pendulum back and forth along a linear path, and the second to observe. But even if you find it necessary to be both observer and experimenter, you should be able to see the basic effect.

Although Pulfrich is given credit for having originated the illusion that bears his name, he was never able to see it himself since he was blind in one eye. This is all the more remarkable because it is his theory that continues to be the most widely agreed upon account of the illusion. His explanation, in turn, is one that incorporates a visual mechanism akin to retinal disparity.

When retinal disparity was discussed in the last chapter, depth
was said to arise from the integration of disparate points on the
retinas, i.e., slightly different images formed on corresponding
points on the two retinas were ultimately seen as having depth.
Such a disparity can be thought of as a disparity between relative
locations on the two retinas. Pulfrich theorized that the effect of
the filter in the pendulum demonstration was to delay, ever so
slightly, the time it takes for the neural "message" to travel from
eye to brain. The implication of this is that it takes longer for the
dimmer image's message to get to the brain than it does for its
more intense counterpart. Normally this delay in transmission
time for the dimmer image's message would be a trivial matter
since the delay is no more than a few thousandths of a second and
also because it is ordinarily the case that the retinal images for
both eyes are of equal intensity (for example, when you wear sunglasses, the neural messages from both eyes are equally delayed).
But when one eye's impulse is delayed relative to the other eye's,
a disparity is created. Since the lagging representation is of the
image when it occupied a different spatial position, the temporal
lag creates a spatial disparity at the level of the brain. Thus the
temporal lag produces spatial disparity only for moving objects,
and only moving objects exhibit Pulfrich's effect.

The hypothesis that temporally disparate inputs to the brain
give rise to depth perception has found considerable experimental
support. If Pulfrich's account is essentially correct, then certain
conclusions automatically follow from it. One consequence is that
an observer should require more time to react to a dim stimulus
than to the very same stimulus seen under higher illumination.
Sure enough, carefully controlled tests of reaction time have
shown repeatedly that this relationship exists.

While most modern reaction-time experiments require electronic instrumentation, you might like to test the intensity–transmission time principle using a much older (and easier) technique.
The only equipment required for your experiment is a yardstick,
but you will need one other person's assistance. Here's how it
works: one person must serve as observer while the other con-
ducts the experiment. The yardstick is suspended vertically by the

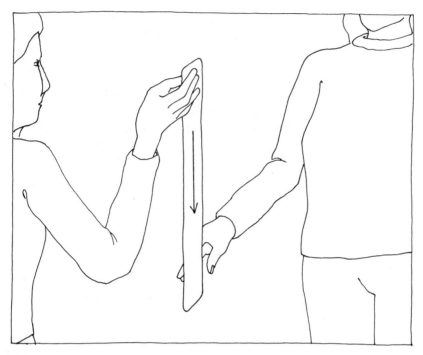

A simple reaction-time experiment. The location of the subject's fingers when he catches the falling yardstick indicates how long it takes him to respond.

experimenter so that it hangs between the thumb and forefinger of the observer's hand, but it must not actually touch the fingers. When the experimenter releases the yardstick the observer attempts to catch it as quickly as possible; the longer it takes him to react, the greater the length of stick that will pass between his fingers. If the fingers are initially poised over the zero point of the yardstick, the numbers then serve as a measure of reaction time as well as distance.

This technique for measuring reaction time is not terribly precise, so many repetitions of the task are necessary in order to arrive at an average value. So as to minimize the effects of anticipation, the experimenter must release the stick at various times as the task is repeated, sometimes quickly after both participants are in position, sometimes at longer intervals. It would be a good idea to perform at least 50 repetitions and then obtain a mean (average) reaction time from these measurements. Assuming that these trials were done under normal room illumination, repeat the entire procedure with reduced illumination (dim the room lights or have the observer wear sunglasses). If all goes well, the second average reaction time should be greater than the first.

Since observers can be expected to improve with practice, you may want to give some practice trials to begin with, or else run a sequence of 100 trials, 50 under normal illumination and 50 under reduced illumination, but in a random sequence. Careful record-keeping will allow you to distinguish the two conditions when you analyze the data. It's also a good idea to let the blank side of the ruler face the observer and the scaled side face the experimenter since knowledge of results can also affect reaction-time scores.

The principle that dimmer images take longer to affect the brain has an additional implication. There is a regular relationship between image intensity and reaction time, such that dimmer and dimmer images result in correspondingly slower reaction times. This relationship, which also has strong experimental support, means that the less intense the image to one eye in the Pulfrich illusion, then the greater the disparity will be, and therefore the greater the depth one will see in the pendulum.

Controlled experiments of these principles are difficult for the casual experimenter to attempt, but an informal proof can be seen by noting the extent of the curvature in the pendulum's path when it is viewed through increasingly dense filters held over one eye. To make the comparisons legitimate, it is best to observe a pendulum that moves at a constant speed, such as the arm of a mechanical metronome or the pendulum of an old-fashioned clock. While it might be possible to have someone swing a pendulum in a regular fashion, this is not the most reliable method. By stacking filters one on top of the other, you can decrease the amount of light they transmit, thereby increasing their density. It is the densest filters, short of blocking light entirely, which would be expected to create the most circular orbits observed in the swinging pendulum.

The speed of the pendulum has also been shown to affect the observed circularity of the orbit, the relationship being that a faster pendulum increases temporal disparity and therefore increases the apparent circularity of the pendulum's path. Test this concept by using a single filter but varying the speed of the pendulum. No special equipment other than a pendulum should be necessary.

One other implication of Pulfrich's theory is that the depth created by the differential illumination of the eyes should occur for many kinds of stimuli in addition to the pendulum, as long as they are laterally displaced. An easily observed application of the Pulfrich principle can be obtained by filtering one eye while looking at the passing landscape from the passenger's position in a moving vehicle. James T. Enright, who originated the demonstration, reported that not only is distance distorted in the demonstration, but size as well, undoubtedly because of the intimate relationship between size and distance.

Enright's phenomenon is much like the Pulfrich pendulum. An observer watches the passing scene from the passenger's position of a slowly moving vehicle while one eye is filtered.

Enright also found that a motion picture taken from the passenger's position is a sufficient stimulus to produce the distortions. Members of an audience who viewed the film all wore sunglasses with one glass removed and saw the same distortions that actual passengers had witnessed. The film was found to be most effective when both nearby and distant objects were in view, and when the speed of the automobile was about 10 miles per hour. Occasionally the effect can be seen in television programs watched with differential filtration, but since television camera work is often static, the Pulfrich effect can be seen intermittently at best. A more reliable way to produce the illusion is to watch moving automobile traffic at a busy intersection, again with one eye filtered. The distortions of size and distance are most striking with the filter worn over the left eye, according to Enright.

In all of the Pulfrich effects it's a good idea to contrast the effect with the same scene viewed without a filter, to remind yourself how extraordinary the Pulfrich pendulum really is.

14. The Anamorphic Puzzle

Those of us who study visual perception are fundamentally interested in the question of why the world looks the way it does. This seemingly naive problem is, upon closer inspection, maddeningly complicated. Let's consider a fairly straightforward component of vision, the perception of shape. Why does one object appear to be circular, another square, and so on? This is not a question of verbal labels, which are, after all, learned. It is more a question of how we are able to discriminate among the myriad of shapes all human beings encounter throughout their lives. Any answer to the problem must first take into consideration an even more fundamental problem, that of *shape constancy*. Examine the first illustration, a photograph of a perfectly ordinary hallway. What shapes are the windows, the doors, and the pictures on the wall? Most people would say that they are rectangular or square, as the case may be, and a geometric measurement of the objects in their natural setting would confirm these judgments. The fact that our judgments of the objects' shapes correspond to their physical constitution is certainly not surprising, but it might be if the retinal image were more closely examined.

Therefore, let's ask a slightly different question: What is the shape of the image that one of those objects casts on to the retina? The second illustration shows some of the photographed objects in isolated outline form. Since both a photographic print and the

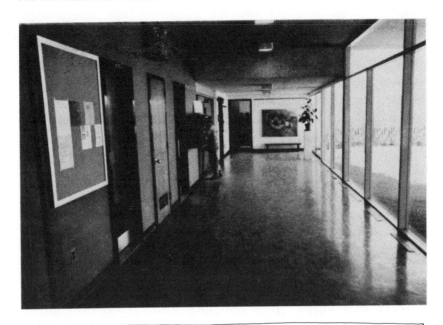

Shape constancy: a photograph of an ordinary hallway. What shapes are the windows, doors, and bulletin board? The outline drawing emphasizes the shapes of these objects as they are projected onto the retina. The shapes in the retinal image are trapezoids, not squares and rectangles, yet we have no difficulty in correctly identifying the true shapes of the objects.

retinal image are optically projected displays, it follows that the outline of images on the retina resembles those in the photograph. How is it that the trapezoids of the photograph are regarded as squares and rectangles? As it happens, whenever we view an object at a slant, the shape of its retinal image differs from the measured shape of the object. So we are left with a predicament. One's judgment of the shape of an object remains constant despite changes in the shape of the retinal image. Why should this be? If shape were determined by the shape of the retinal images alone, then objects' shapes would continually alter as we moved about.

The general answer to the problem of *shape constancy,* as the relationship is known, is that in addition to the shape of the retinal image, we also seem to take the slant or orientation of the object into consideration. Both retinal image shape and the distance to points on the object contribute to our ability to maintain stable impressions of shape. (Note that these two variables are highly reminiscent of the retinal image size and distance factors that maintain size constancy.) The two-factor theory of shape constancy has been confirmed in experiments in which subjects were asked to describe the shape of a slanted object with and without distance information present. In the absence of distance information, and therefore without an indication of slant, subjects frequently judged the shape of the object to be consistent with its retinal image shape instead of the objective shape. When this happens, neither size constancy nor shape constancy is evidenced. Given such experimental viewing conditions, someone could readily say that the bulletin board or a window frame in the photograph looked trapezoidal.

This brief introduction to shape constancy creates a problem of its own, one which makes the original question that much more complicated. Once someone understands the basis of shape constancy, it is possible for him to distinguish between a judgment based on shape constancy and one based on how the retinal image is thought to look. To return again to the first illustration, if I now asked you what shape a window is, you might in all candor answer either that it is rectangular or that it is trapezoidal, depending on how you interpret my question. It is precisely this kind of subtlety that gives researchers gray hairs.

Although we treat objects as being consistent with shape constancy in everyday life, there are a few instances when we do not, cases in which the correct retinal image shape must be obtained before shape can be accurately evaluated. We could say that shape constancy has limits and that extreme alterations in the shape of the retinal image are increasingly more difficult to interpret. The complex interaction of the variables contributing to shape constancy is nicely illustrated in anamorphic art.

The anamorphic picture is a type of drawing or painting in which a scene is projected onto a tilted, curved, or otherwise distorted surface. The technique first appeared in the fifteenth century and persisted well into the eighteenth century, finding favor as a visual puzzle and amusement. The examples shown here are all of the same design, projections of pictures onto tilted surfaces. Perhaps it would be helpful if you were to think of them as stretched-out versions of normal drawings. When viewed as ordinary pictures they look quite peculiar, but they suddenly become recognizable when viewed from the side of the page. For some reason, shape constancy does not work for these drawings when they are seen as ordinary pictures. An explanation of the failure of shape constancy in anamorphic art may be that the second component in the shape constancy equation, distance, is absent, or at least ambiguous, in these works. Remember that in the absence of distance cues our judgments of shape are based largely on properties of the retinal image.

The two anamorphic pictures, drawn by an anonymous artist around 1868, were intended to be viewed almost on edge, the effect of which is to project an image on the retina that approximates the objective shape of the characters depicted in the drawings. When the drawing is viewed on edge (with one eye only), the page must be held quite close to the eye, thus effectively negating depth by convergence and stereopsis.

That depth information is critical to accurate shape perception is informally illustrated in the next pair of pictures. In the first

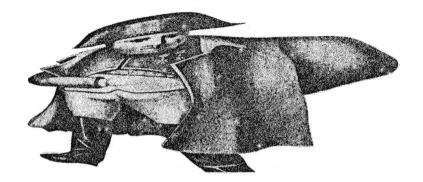

Anamorphic drawings. Hold the page nearly on edge and view them from the positions marked.

From F. Leeman, J. Effers, and M. Schuyt, *Hidden Images: Games of Perception, Anamorphic Art, Illusion.* Published in 1976 by Harry N. Abrams, Inc., New York, N.Y. Copyright © 1975 by Verlag M. DuMont Schauberg, Cologne. Reprinted by permission.

photograph sufficient indication of depth is provided by pictorial depth cues (present in the background) to make the condensed image of the billboard readily identifiable as a rectangular shape containing a normally proportioned picture. Once the depth is made unclear, as was done for the next photograph by removing the background, the billboard takes on the distorted appearance of an anamorphic picture. Because the distortion is a condensa-

The importance of distance information to shape constancy. The shape of the billboard is rectangular when distance cues, provided by the background, are present. The same billboard in isolation has the quality of an anamorphic drawing.

The elongated shadow of a pair of scissors is anamorphic in this photograph. It will appear normally proportioned when the picture is viewed
on edge.

tion in two dimensions, it cannot be made to look rectangular by
just tilting the page and viewing it on edge. The last photograph,
however, is a record of the elongated shadow cast by a pair of scissors (the scissors, photographed from directly above, appear as a
dark line at one edge of the shadow). The stretched image, like
those of the anamorphic drawings, can be made to look more normal by viewing the illustration from the side.

15. The Probability of Form

Any discussion of shape perception, also called the perception of form or pattern, would be incomplete without some mention of *ambiguous forms*. Under ordinary circumstances the external world has a certain permanence or reliability, a quality that transcends shape constancy. If you see a book or a face, it not only retains its shape as you change your position relative to it, it also maintains its identity. In other words, the face is not normally transformed into a book or vice versa. While this is the usual way of things, there are exceptions, occasional instances when the very identity of what you see may vacillate. What is particularly intriguing about these ambiguous forms is that the alternative ways of seeing them are lawful, they are seen pretty much the same way by everyone.

The descriptive labels attached to these forms tell us something about how they are typically seen, as well as the ways people have theorized about them. In addition to being called ambiguous, the forms have also been known as *reversible, equivocal, improbable,* and *multistable*. Not all of these terms are interchangeable, of course, but all refer to visual displays, usually pictures, that elicit from the observer two or more clearly different interpretations. These different interpretations are wholly predictable, and are nothing like the idiosyncratic hallucinations that fascinate psychiatrists. Yet, so curious are these pictures that the uninitiated

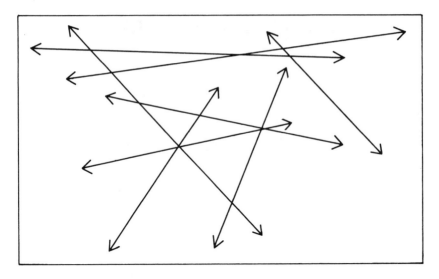

Ambiguous forms need not be complicated. This pattern of arrows, created by J. O. Robinson, can be seen to occupy many different orientations in space.

From J. O. Robinson, *The Psychology of Visual Illusions*. London: Hutchinson, 1972. Reprinted by permission.

observer is likely to regard them with a mixture of amusement, curiosity, and consternation.

Not all ambiguous forms have been contrived for that purpose. A few dots or lines may be all that are required to create the effect, as you can see from the example presented here, or by cre-

Even a line scribbled through a circle will create two complimentary forms. Try your own in the two unfilled circles.

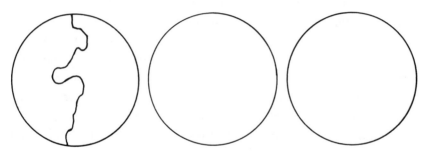

ating a few patterns of your own. Fred Attneave has noted that any scribbled line drawn through a circle will create two coherent but alternating forms. The ambiguity of sparse line drawings was used to advantage by the entertainer Roger Price in the 1950s in a genre of comical ambiguous drawings that he called *droodles*. The droodle that I've included here has a number of alternative interpretations, such as a farmer walking past a window, or a jet-propelled "W." Feel free to add your own titles.

Repetitious patterns are also effective equivocal forms. A matrix of dots, triangles, or any other geometric forms can be seen in a variety of ways. There seem to be two sub-types, which I will re-fer to as *tessellated* and *untessellated*. The tessellated pattern is composed of interlocking elements, like the tiles of a mosaic, while the stimuli of untessellated patterns are less likely to be seen as bordering on one another. The white dot pattern is of the sec-ond type and will fluctuate continuously in its appearance, pro-ducing at different moments impressions of hexagons, parallelo-grams, circles, and so on. Tessellated patterns are apt to have an additional quality, a property the Gestalt psychologists called the *ambiguous figure-ground* relationship, a concept that deserves ad-ditional consideration.

A droodle.

The reputations of several of the Gestalt psychologists centered upon their insightful studies of form perception. Edgar Rubin, a Danish psychologist, made a fundamental distinction when he noted in a 1915 paper that whenever we observe our surround-ings, we tend to regard some portions of the visual world as fig-ure, while the remainder is ground. The figure is the form, the delineated and tangible aspect of the scene, while the ground is akin to a background; it is poorly delineated, more amorphous than the figure, and is seen to be "behind" the figure. The distinc-tion is better understood by examining ambiguous figure-ground drawings, such as Rubin's own reversible vase-profile drawing. From time to time the viewer will see either a pair of black pro-files against a white ground, or a white vase set against a black background. Rubin's illustration is tessellated because the border of the vase exactly coincides with that of the profiles. The two al-ternatives will reverse spontaneously as you examine the picture.

At any given moment the pattern of white dots will be spontaneously organized into geometric forms.

From M. Luckiesh, *Visual Illusions: Their Causes, Characteristics and Applications*. Copyright © 1965 by Dover Publications, Inc. Used by permission.

The repetitive triangles are tesselated; the borders between the black and white triangles are shared, much like the interlocking of mosaic tiles.

It is of interest to note that in Rubin's illustration the viewer sees the black portion as figure or the white portion as figure, but not both simultaneously. The boundary between the black and white areas also attaches to one figure at a time and cannot be shared by both.

By carefully studying Rubin's and other uncomplicated drawings, Gestalt psychologists attempted to understand the factors that favored the perception of figure over ground. By using drawings like reversible Maltese crosses, it was found that the orientation of the figure as well as the relative extent of different areas within the picture contribute to the determination of form. Literally dozens of these organizing principles were set forth in addition to orientation and area. The ambiguity of figure and ground found in Rubin's modest drawing is taken to a higher level of elaboration in the graphic works of the Dutch artist Maurits C. Escher. Even though Escher was a contemporary of many of the Gestalt researchers, there is no convincing evidence that his artistic accomplishments were directly influenced by them.

Escher's unending curiosity with interlocking forms was more likely the product of a visit that he made as a young artist to the Moorish palace known as the Alhambra, located in Granada, Spain. He commented years later upon the extraordinary mosaic tiling that he had seen there:

> This is the richest source of inspiration that I have ever struck; nor has it yet dried up. . . . A surface can be regularly divided into, or filled up with, similar-shaped figures which are contiguous to one another, without leaving any open spaces. The Moors were past masters of this. They decorated their walls and floors . . . by placing multi-coloured pieces of majolica together without leaving any spaces between. What a pity that Islam did not permit them to make "graven images."

Escher labored under no such prohibition, and his graphic works displayed an unrivaled combination of technical skill and ingenuity. His tessellated graphic works sometimes employ repeated figures, sometimes not, but they always produce the reversibility of figure and ground first described by Rubin.

Rubin's famous profile-and-vase illustration showed the properties of the figure-ground relationship.

Reversible figures of the Rubin and Escher type may well involve the operation of depth cues, the pictorial cue of interposition in particular. Rubin, it should be remembered, said that the ground extended behind the figure, a fact that you can verify by re-examining the vase-profile picture. The figure always appears to be on top of the ground, whether it is the black portion or the white portion that is seen as figure. I find it of interest that every ambiguous drawing that I have seen seems to involve a reversibility in the third dimension. Researchers like Richard Gregory and M. H. Pirenne favor the view that all depictions of the real world in two dimensions (drawings, paintings, photographs, and the like) are inaccurate. The reason for this is that a conflict among depth cues always exists for the viewer of a drawing. Pictorial cues dictate that a scene in depth exists, while physiological

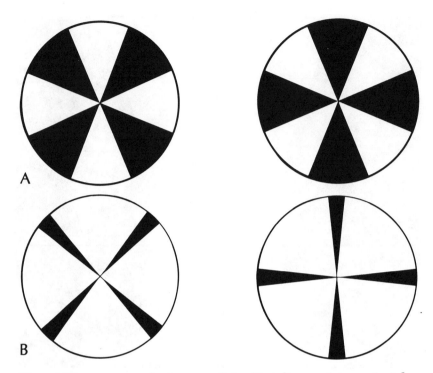

These Maltese crosses depict two of the Gestalt organizing principles:

A. In this pair, the white cross and the black cross, respectively, tend to be seen as figure. This suggests that there are favored orientations for the perception of form; in this case, the vertical.

B. The smallest enclosed area (the black cross) is also favored to be seen as figure. The principle is not absolute since it is also possible to see a white cross in the same illustration.

depth cues (notably stereopsis) emphasize that the drawing is, after all, flat. In order to see the painted, drawn, or photographed depiction as being in depth, one must at some level disregard those depth cues that make it seem two-dimensional. It is therefore not surprising that pictures are more likely to harbor ambiguity than is the real world. This is doubly so because the depth portrayed by pictorial cues has a learned foundation, and is to

---

The sophisticated use of the figure-ground principle is seen in Escher's woodcut *Day and Night* (1938).

Maurits-Coenelis Escher, *Day and Night*. Yale University Art Gallery. Gift of George Hopper Fitch, B.A. 1932, and Mrs. Fitch. Reprinted by permission.

M. C. Escher's graphic works often manipulated figure-and-ground. These sketches of tiling at the Alhambra were drawn by Escher in 1936 and were an important influence upon his later work.

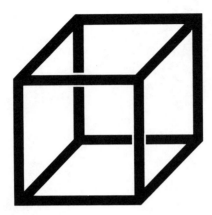

The ambiguity of the Necker cube is attributable, in part, to spatial factors. When the overlapping of lines is made more distinct, the cube is less likely to reverse its apparent orientation.

some extent a convention of artistic representation rather than a law of nature. Thought of as rules of artistic representation, pictorial depth cues, in the absence of physiological cues, are to an extent arbitrary and malleable. I would therefore argue that the viewer's confusion in interpreting these figures is a confusion of spatial localization since information about depth is not powerful enough to favor one interpretation of figure over another.

The spatial factor is more obvious in other kinds of ambiguous figures, such as the Necker cube you looked at in the discussion of completion. The Necker cube is ambiguous, in part, because one cannot determine from the drawing itself whether components of the cube are above or beneath one another at the points where lines intersect. This being the case, the cube has equal probability of existing in either of two spatial orientations. The ambiguity can be reduced by redrawing the figure in such a way that the overlapping of the lines is made clear. The same principles apply to many skeletonized outline drawings, not just the Necker cube. The fact that the drawings with clear overlapping can still be reversed, although only with considerable difficulty, merely points out the important role normally played by stereopsis in the interpretation of the solidity of objects.

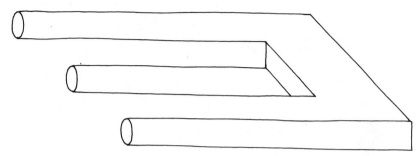

Impossible figures are drawings in which two parts of the picture appear in incompatible perspective.

Impossible figures are a class of ambiguous figures that deserve special mention. They are drawings in which two parts of the picture have been drawn in incompatible perspective. At first glance such a drawing appears normal, but closer inspection shows that while isolated portions appear to be in correct perspective, these elements perpetually defy integration.

Many geometric outlines show reversal, including this tetrahedron.

From M. Luckiesh, *Visual Illusions: Their Causes, Characteristics and Applications.* Copyright © 1965 by Dover Publications, Inc. Used by permission.

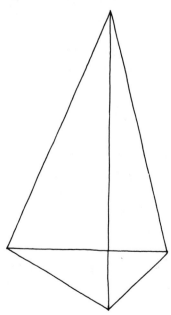

Cochran's "Freemish Crate" makes the point that figures are not so much impossible as they are improbable. The object created for the photograph looks impossible only in one orientation. At other viewing angles it appears to be a collection of oddly jutting struts connected at many different angles.

"Freemish Crate" by C. F. Cochran. Used by permission of C. F. Cochran.

Richard Gregory has pointed out, quite correctly, that these pictures are not so much impossible as they are improbable. The term "impossible" in the present context implies that it would be impossible to make a three-dimensional object that exactly corresponds to the drawn object. C. F. Cochran, among others, has done exactly that—created, in a photograph, objects that appear to be impossible. The photograph must be taken at just the right angle in order for the demonstration to succeed, for at any other vantage point the object appears to be a twisted framework bearing little resemblance to the regular figure in the photograph.

Recent research conducted by Thaddeus Cowan and Richard Pringle at Kansas State University has shown that drawn figures exist along a continuum of likelihood that can be ordered from the highly probable to the highly improbable. Their subjects ordered 27 pictures along this dimension with considerable agreement (the pictures were generated by a specially designed computer program). They found that the judged likelihood of a figure

Cowan and Pringle's ordering of a figure in terms of likelihood. The first drawing is most probable, while the twenty-seventh is least probable.

From T. M. Cowan and R. Pringle, "An investigation of the cues responsible for figure impossibility," *Journal of Experimental Psychology: Human Perception and Performance*, 1978, 4, 112–120. Copyright 1978 by the American Psychological Association. Reprinted by permission.

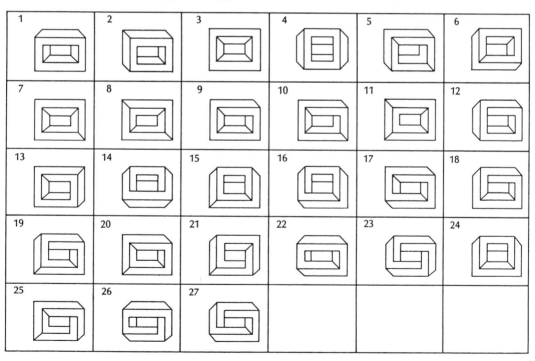

has to do with the handedness or *parity* of the twists that portions of the figure describe, and this variable in turn may be related to how portions of the figure are seen in depth.

Our growing understanding of ambiguous and improbable figures has helped to temper our uncritical trust in the fidelity of forms portrayed in two dimensions. While everyone acknowledges that pictures may sometimes "lie," only now are we beginning to understand why. Rather than saying that pictures lie, it might be preferable to say that these two-dimensional representations of events in the third dimension exist along a continuum, at one end of which are the Necker cube and Cochran's freemish crate, and at the other, graphic works made with such consummate skill that the spatial discrepancy defies detection.

16. The Restless Eye

As a youngster I possessed an artistic bent that my parents encouraged by sending me off to Saturady morning drawing classes. It was during these pleasant intervals that I first heard a popular belief among artists—namely, that a picture should contain a string of features that would carry the viewer's eye across the paper or canvas. This idea that good artwork creates a "line" attracting the viewer's eye in a predictable fashion has been the sustenance of many an art teacher and art critic alike. Is it true that a painting has good composition because it causes the eye to move in certain ways? On a more fundamental level we might ask how the eye objectively moves in response to the environment and the significance of eye movements to the understanding of form perception.

The study of how our eyes move has long been a topic of interest to vision researchers, possibly because eye movements are so accessible. Unlike investigations of the interior of the eye or the visual nervous system, eye movement research requires no surgery or biological paraphernalia. At the heart of these investigations is the notion that systematic relationships may exist between the way a person's eyes move and the way that person sees the world. Perhaps, it is reasoned, the eye movements actually cause us to see the world the way we do. This principle has been applied not only to theories of art appreciation, but to the study of reading and form perception as well.

DISCONTINUOUS FIXATION

First, let us examine the ways that the eyes move, and why. Movements of the eyes are controlled by three pairs of opposing muscles attached to the outside of the eye. In the normal person,
sleeping or awake, the eyes are ceaselessly in motion. You are already familiar with some eye movements, such as the convergence
motions that the eyes make in order to maintain an image in
foveal vision as the object approaches the viewer. The eyes also
make voluntary tracking motions, for example when you follow

Involuntary eye movements. The path of the fovea is recorded as an
observer attempts to fixate a stationary object. Three types of eye movements are evidenced: slow *drift* movements, upon which are superimposed *tremor* movements, and the rapid, straight-line *saccadic eye
movements*.

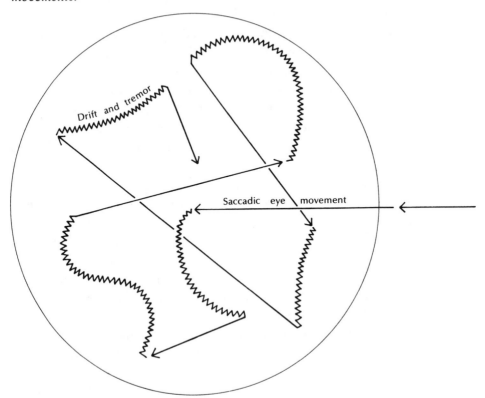

the position of the ball at a tennis match. Tracking movements also maintain the image in the general area of the fovea, that small landmark on the retina that is responsible for the best visual acuity.

There are, in addition, several involuntary eye movements that are so subtle in their operation that the observer is rarely aware of them. Suppose that you set out to "look at" (fixate) a stationary object and tried to keep your eyes as motionless as possible. Even in this instance your eyes would exhibit three types of involuntary eye movements: If, initially, the object had been imaged on or about the fovea, we could expect to see a slow *drift* movement away from the original position. Superimposed on the drift would be a very rapid *tremor* motion as well. At some point in the drifting motion there would occur an extremely rapid compensatory flick of the eyes to bring the image back into the area of the fovea. These rapid, straight-line movements of the eye are called *saccadic eye movements*. In the course of fixating the object, there would be a continuous alternation of drift and saccadic eye movements as long as the observer stared at the object. The visual system responds to flux rather than to a steady state, and because of this principle involuntary eye movements should seem perfectly reasonable and natural to you.

Let's return to the problem of composition. How do the eyes move when confronted with forms? Lloyd Kaufman and Whitman Richards examined one aspect of this problem, the results of which may surprise you. They were interested in how the eyes behave when viewing relatively small objects, and they used a research technique that permitted them to periodically plot where on a picture a subject was fixating at a given point in time. By repeatedly sampling this behavior they were able to determine the fixation tendencies of subjects for several simple patterns. In one variation, the subject was asked to plot where on a line drawing he thought he was looking, and this was compared with an objective measure of fixation. There was a considerable discrepancy between the two measures, a fact which should surprise those who believe that we are capable of judging the position of our own eyes with accuracy.

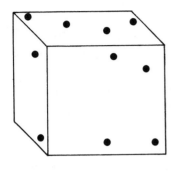

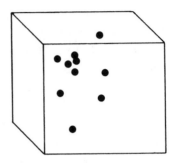

The differences between actual fixations (below) and where the observer believes his eyes are pointing (above) can be substantial.

From L. Kaufman and W. Richards, "Spontaneous fixation tendencies for visual forms," *Perception and Psychophysics,* 1969, 5(2), 85–88. Copyright 1969 by The Psychonomic Society, Inc. Used by permission.

Kaufman and Richards found that for a simple drawing of a cube, the location most frequently fixated was at the center of the drawing; central fixation was also the case for other small, geometric forms. Only for relatively large figures was a tendency found to scan the figure. These researchers showed that we do not perceive small shapes because eye movements trace an outline of the pattern, a fact that you can readily verify for yourself.

In the laboratory, the measurement of continuous eye movements necessitates the use of complex and expensive equipment, such as cameras and electrophysiological recording instruments. Less complicated techniques exist for measuring discontinuous fixation tendencies (where your gaze is directed at a given moment rather than an uninterrupted record of where the eyes are pointed continuously). Even though measurements of fixation tendency lack some of the elegance of continuous recordings, they are inexpensive and capable of telling us a great deal about eye movements and perception.

Several of the techniques commonly used to measure discontinuous fixation tendency take advantage of *entoptic visual phenomena*. These phenomena produce sensations of vision, but the sensation originates within the eyeball rather than in the outside world. When, for example, you press lightly against a closed eyelid and "see" a spot of light, a common phenomenon called a *phosphene*, you are witnessing one kind of entoptic phenomenon. Yet another entoptic phenomenon is the *Purkinje tree:* While staring at a blank wall or paper, hold a small flashlight near the outside corner of one eye and gently shake the flashlight up and down. As you do so, you will be aware of a faint, treelike image. It is not "out there," but a pattern of shadow formed by the blood vessels on the interior of the eye. It is even possible to make a sketch of the pattern as you continue the illumination. Entoptic phenomena not only illustrate properties of the eye, but can also be used to mark where the eye is pointed.

Kaufman and Richards used an entoptic marker in their studies, but since some equipment would be necessary to reproduce their particular technique here, I've substituted an even easier entoptic phenomenon, *afterimages*, for the next few demonstrations. In the chapter on size perception (Chapter 11) you used an afterimage to illustrate the properties of Emmert's law, but here it will be used only as a reference of retinal location. Since the afterimage is due to the fatiguing of the retinal receptors themselves, what you "see" in the afterimage is an indication of how the retinas are positioned in relationship to the outside world. Because the afterimage is most likely to be formed on the fovea, it shows the direction of gaze with reasonable precision.

In all of the following demonstrations, you will first have to form a discrete afterimage, and I recommend that you do this by staring at a moderately intense light source, like a shaded lamp, over which has been placed a paper mask with a small circle removed. If you stare at the bright area for a half-minute or so at a distance of a few feet, a strong afterimage will be created that will endure for several seconds after you redirect your gaze.

The first demonstration is to note where the afterimage appears on simple geometric figures, like those used by Kaufman and

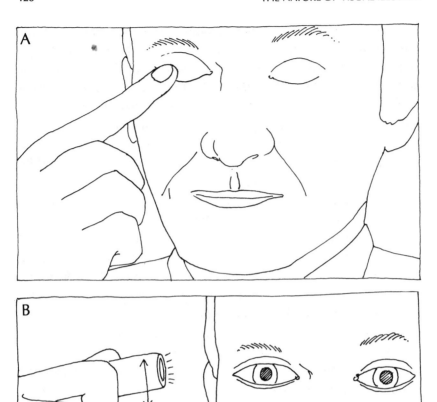

Examples of common entoptic phenomena:

A. Phosphenes can be experienced when you gently press against a closed eyelid.

B. The Purkinje tree is the observation of the pattern of blood vessels within your own eye. As you stare at a blank wall, rapidly oscillate a flashlight aimed at one corner of your eye.

C. Afterimages are also considered entoptic phenomena. A distinct afterimage, like that obtained from looking at a small light source, can serve as a convenient marker of the eyes' position.

Richards. While this is most definitely not a carefully controlled experiment, I think you'll find that the afterimage, and therefore the fovea, tend to be positioned more toward the centers of the figures than along their borders.

Some theories have proposed that the way we see ambiguous figures and drawn illusions is a function of eye movements. One obvious disproof of the theory as it applies to ambiguous figures is to use the afterimage as a way of maintaining your gaze (thereby minimizing eye movements), and then seeing if reversibility still occurs.

As one example of the technique, try the following: Select an ambiguous figure from the last section and prop it up in front of you at reading distance. Once again form an afterimage, but now try to keep it, and therefore your eyes, in one location on the figure, perhaps at the center of the drawing. You should see that even with your eye movements minimized, reversibility can nonetheless occur. This is true even if your eyes are pointed at a different feature of the figure, such as an outside corner or line. The demonstration underscores the point that the reversibility of ambiguous figures can not be attributed to eye movements. The eyes do not move one way with the Necker cube seen in one orientation and a different way when the alternative orientation is seen.

Repeatedly form afterimages and note where your eyes fixate these common geometric forms. The precise method for forming afterimages is given in the discussion of size perception.

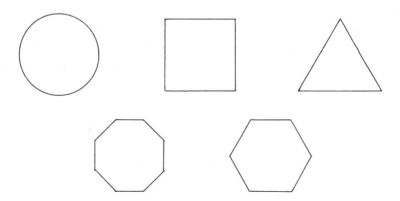

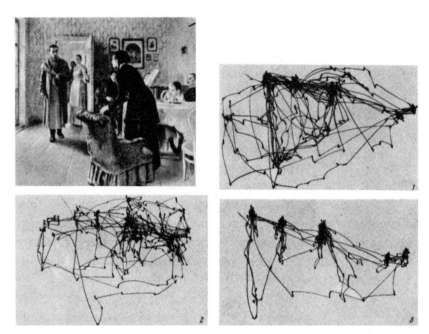

Soviet researcher A. L. Yarbus has shown how the pattern of eye movements can change as a function of instruction. Subjects were asked to
examine I. E. Repin's picture "An Unexpected Visitor" for a three-
minute interval. The eye movement records are for single subjects
asked to: 1. simply examine the picture (free inspection); 2. "estimate
the material circumstances of the people in the picture"; or 3. give the
ages of the people portrayed in the picture. The eye movement patterns mirror shifts in attention.

From A. L. Yarbus, *Eye Movements and Vision*. Copyright © 1967 by Plenum
Publishing Corp. Reprinted by permission.

You may find it instructive to see if eye movements contribute to
the ambiguity of the other reversible figures as well. The clear implication of all this is that the reversibility of ambiguous figures
occurs at a stage beyond the eyes, somewhere in the central
nervous system. Several other lines of evidence agree with this
conclusion.

Of course we do not routinely view small objects exclusively.
The subject matter we ordinarily see extends well beyond the narrow limits of the fovea and requires a sequential movement of the

eyes, a kind of sampling over time. Might it still be possible that, for artwork and other large displays, the pattern of eye movements dictates something about the forms we see?

When someone examines one of these large displays, the eyes advance over it in a series of saccadic eye movements. Because the image is blurred during the movement, and perhaps for additional reasons of physiology, it is extremely difficult for the viewer to see while the saccadic movement is happening. For all intents and purposes, then, we have no vision while the movement is occurring, only at its terminating points. Even if the eyes traveled a path that exactly mimicked the features of the object under observation, only a fraction of the retinal events would be transmitted to the brain. The suppression of vision during a rapid excursion of the eyes can be experienced by your forming an afterimage and then watching what happens to it as you deliberately flick your eyes about. You should see the afterimage at the start of the movement and at the end, but not during the actual excursion of the eyes. Any hypothesis about the relationship between eye movements and form perception must account for the integration of these discrete visual samples over time.

What is there that is consistent about the pattern of eye movements that is recorded when one is given an extended period of time in which to examine a display whose dimensions extend beyond the confines of foveal vision? For one thing, under these circumstances people do not look at the display in an entirely random fashion. The eyes tend to rest on angular features far more than on homogeneous areas (but the sequence of eye movements from feature to feature is different for different observers), and in recognizable pictures more time is spent looking at colored regions and faces, should they be present. The Soviet researcher A. L. Yarbus has also shown that the intention of the viewer examining a picture can profoundly influence the pattern of eye movement over time. Given different instructions about what to look for, observers of the same picture produce dramatically different patterns of eye movements. Therefore it can be said that what the artist does can most definitely influence how someone's eyes move about the artwork during prolonged inspection. What is far less clear is the causal nature of events during the inspection. Does the viewer first perceive the form and does then an event in the central nervous system cause the eyes to move about the picture? Or does the pattern of eye movements dictate to the central nervous system the pattern that will eventually be perceived?

COGNITIVE MAPS

Research tells us that patterned eye movements are not absolutely necessary, even for the apprehension of large displays. Some years ago, Theodore Parks created a demonstration that he dubbed *Passing a camel through the eye of a needle,* and which you can reproduce from the materials provided. With the card containing the narrow slit held stationary, pass the outline figure of the camel beneath it. Even though your eyes are prevented from moving about the outline form, you will nevertheless have no difficulty in identifying the form as a camel. To make doubly sure that your eyes can not move freely about the form, you may want to use an afterimage marker to confirm that your eyes fixate the center of the slit.

The Parks effect: passing a camel through the eye of a needle. Cut out and mount the screen (be sure to remove the slit) and the card with the camel outline. You should have no trouble identifying the figure when it is moved horizontally beneath the slit. It is difficult to imagine what role eye movements might play in this example of form perception. See how other outline forms work as well.

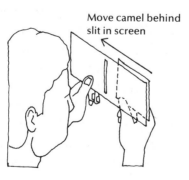

Move camel behind
slit in screen

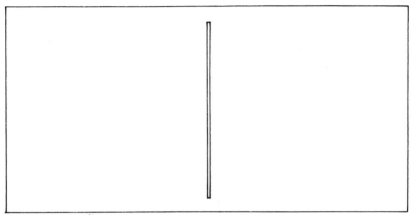

Screen (remove slit)

In both the normal pattern of eye movements and the Parks effect, portions of the stimulus are exposed to vision over time. The psychologist Julian Hochberg has theorized that there is some underlying template or *cognitive map* into which the successive samples are integrated. In the absence of the underlying map, the successive temporal samples make little sense, a point that has experimental support. For example, subjects shown a series of projected images of the foveal samples, one after another, but not told the relative location of one sample to the other, cannot construct a coherent form from all of these bits and pieces. In similar fashion, Irvin Rock and Fred Halper, then at Rutgers University, showed much the same thing for the Parks effect. If a subject is unaware of the fact that the form is moving beneath a slit, and merely observes what then appears to be undulating lines, no coherent form perception results.

What does all of this mean? Apparently the image samples obtained in examining a picture do not in and of themselves dictate the perception of form, whether this is for images within the fovea or for those that extend beyond the fovea. Neither of the patterns of causality proposed earlier is entirely correct. It is not true that eye movements cause form perception or that form perception causes eye movements. Given the limitations in acuity inherent to the human eye, complex form perception requires imaging successive portions of a large display over time and relating the samples to one another within a larger context. No one supposes that there is a literal "map" in your head for that purpose, only that at some level of the brain an integrating function must exist.

17. Bands, Grids, and Gratings

> No object is mysterious. The mystery is your eye.
>
> ELIZABETH BOWEN

Our ability to solve the mysteries of form perception will ultimately depend upon understanding how variations in light intensity alter the visual nervous system. To achieve this goal we must obtain a better idea of the external stimuli that culminate in the comprehension of form. Most people, I think, having never thought much about this subject would be inclined to describe forms as assemblages of lines. Is that an accurate definition?

Imagine the following: A light meter is passed slowly over a sheet of paper on which something has been drawn. The light meter, a more sensitive version of the type found in cameras, measures the intensity of light energy along a numerical scale. As the meter is moved along, the pointer on the scale suddenly deflects and then, just as suddenly, returns to its former level. The abrupt shifts in the intensity of light reflected from the paper might serve as an arbitrary physical definition of a line. Forms, then, could be thought of as a collection of these lines.

Of course a difficulty with defining form as a collection of lines is that not all forms are composed of lines. A filled figure is recognized as a form even though no lines contribute to its existence. Therefore, we must modify our first definition so this stimulus can also qualify as a form. Passing the light meter over the solid form will also cause a sudden excursion of the pointer, but it will take longer than before for the needle to return to its original position.

The common feature in both examples is that the meter shows a sudden excursion, indicating an equally sudden shift in light intensity. It would be better to use a more general term to describe an intensity change, and that term is *contour*. A line is one example of a contour, but there are others, as the filled square demonstrates.

Even though you know better than to talk about lines when you mean contours, it is still not correct to say that form perception involves the organism's response to collections of contours. Why?

A light meter passed over a single line will show an abrupt shift in the intensity of light, a fact also illustrated in the graph.

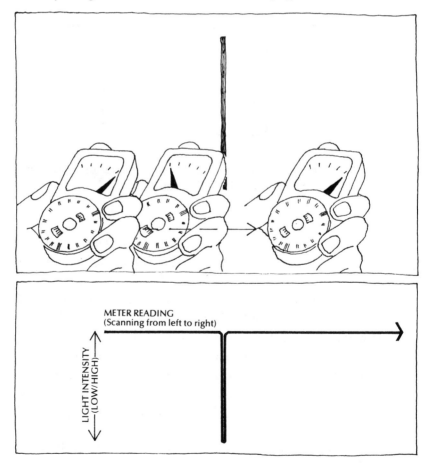

METER READING
(Scanning from left to right)

LIGHT INTENSITY
(LOW/HIGH)

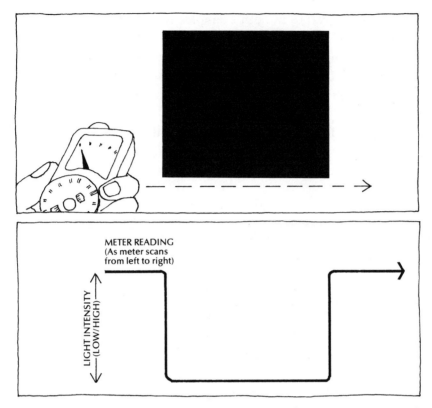

Not all contours are lines, however. If the demonstration is repeated for a filled square, the intensity shift still occurs.

Because there are too many exceptions to this generalization to permit its validity. One exception is the illusory contour figure, some examples of which you saw in the chapter on completion effects. The perplexing aspect of illusory contours is that an edge is seen to exist in areas devoid of physical contours. If the light meter test were repeated for one of these figures, the pointer would not deflect as the meter was moved across the drawing in the region where the illusory figure appeared to float above the elements that induced it. Although the illusory figure looks brighter than its background, the physical intensity of light in both regions is the same.

What you see in these drawings is *not* a quirk. The dark inducing forms cause the illusion to occur, but in ways that are not completely understood. Illusory contour drawings emphasize that the stimuli of form perception remain elusive.

MACH BANDS

Illusory contour figures are not the only figures that pose serious problems for theories of form perception. Consider as well the several *Mach band* phenomena. All are related to observations made by the nineteenth century giant of physics and natural philosophy, Ernst Mach. Many of these demonstrations use rapidly rotating disks to produce circular areas of black, white, and gray. If a disk is half black and half white, it appears gray when spun; increase the amount of white and the shade of gray is lightened proportionately in the spinning disk. In the course of investigating

Yet contours, defined as intensity shifts, are not always a sufficient explanation of form. For example, in this illusory contour drawing, you can see a distinct contour in a region of uniform intensity. A light meter moved along the path indicated by the dotted line would show no intensity shifts.

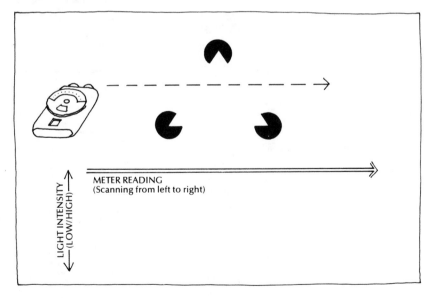

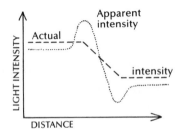

Mach's disk (1865), when rapidly rotated, reflects light in a distribution shown in the graph, yet observers see a distinct light and dark band not predicted in the graph. These Mach bands are attributable to the workings of the visual system itself.

Mach bands are also evident in these stationary patterns that were made to reflect the same distribution of intensities as the disk. Hold the patterns at arm's length and a light vertical line will be seen in the pattern on the left and a dark line in the pattern on the right.

From Floyd Ratliff, "Contour and contrast," *Scientific American*, June 1972. Used by permission of the author.

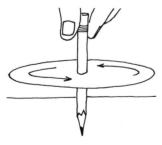

The disk can be rotated by making it into a top: mount on cardboard, insert a short pointed dowel or pencil through the center of the disk and spin it on any flat surface.

this phenomenon, Mach hit upon a peculiar discrepancy between the predicted appearance and the perceived appearance of certain of the disks. If a disk, when spun, reflects an area of uniform high illumination and an area of low illumination, separated by an area of graded illumination, the distribution of light predicts the appearance of a bright area, then a gradually darkening band that blends into a uniform dark area. What actually appears, in addition to the three regions, are two distinct circular bands, one light and one dark. The two additional Mach bands do not correspond to the distribution of light intensities on the disk and therefore must be a product of the viewer's visual system. The rotating disk is not an absolute necessity for producing Mach bands since they can be seen in any display that exhibits the arrangement of intensities of the spinning disk.

Most researchers agree that the appearance of Mach bands is attributable to interaction effects among cells in the visual nervous system. At the retinal level, neurons actively inhibit or excite neighboring cells depending upon how they themselves are stimulated. This process of *lateral inhibition* and *excitation* enhances the contrast between dark and light regions; i.e., it exaggerates, ever so slightly, the contrast at contours.

The Hermann grid may also be an expression of the nervous system processes of lateral excitation and inhibition. Notice the brightness differences at the points where lines intersect.

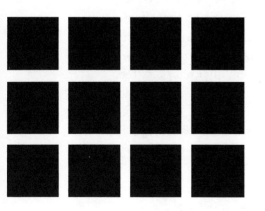
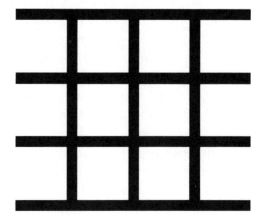

GRIDS

The same interactive process may also account for the *Hermann grid*, a regular matrix of dark squares separated by white paths of uniform width. As you glance over the grid you will be aware of shadowy dark patches at the intersections of the white paths. The best-known theory of the grid holds that the dark squares enhance the portions of the white paths adjacent to them. But at the intersections no such enhancement is possible in the absence of the squares, and so the areas of the intersections appear, by comparison, slightly darker. If the effect is due to interactions among retinal cells, it is more pronounced in the periphery of the retina than at the fovea. Notice that the shadowy patches do not form at the fixated intersections, only at intersections imaged outside of the fovea. If the Hermann grid is caused by lateral excitation of the retina, lateral inhibition should also occur, and indeed, when a pattern of white squares on a black background is viewed, light patches appear at the intersections of the dark pathways.

In all of these demonstrations you have seen something not predicted by the analysis of the distribution of light energy in the stimulus. That is why we need the term *brightness*, since it recognizes that what we see is also a function of our nervous systems, and that perceived intensity is only partly based on the intensity of light.

For the person interested in form perception, life would be reasonably simple if form perception could be limited to the effects of contours and the retinal interactions typified by Mach bands and the Hermann grid. As usual, that is too much to hope for. Other demonstrations produce different but equally unanticipated results. One of my favorite demonstrations also uses a spinning disk, but with quite a different outcome than before. *The Craik-O'Brien effect*, named for the two researchers who discovered it independently of one another, is best illustrated by an illusory disk invented by Tom Cornsweet.

The disk is mostly white, except for a pie-shaped black wedge, along which appear two spurs, one white and one black. Except

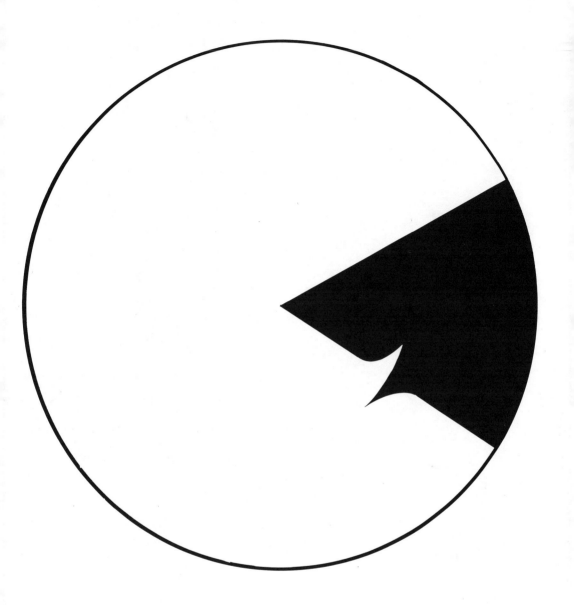

The Cornsweet illusion (Craik-O'Brien effect). Again, the brightness distribution perceived in the spinning disk is not predicted from the distribution of light energy. The two spurs should create a somewhat brighter and a somewhat darker band than the surrounding gray field. Instead, the observer sees two uniform zones.

Richards's disk produces a ring that looks lighter or darker than its background depending on the distance of the viewer from the rotating disk. The phenomenon suggests the importance of spatial frequency factors.

Reprinted with permission of author and publisher from W. Richards, "Illusory reversal of brightness contrast," *Perceptual and Motor Skills*, 1968, 1169–1170, Figure 1.

for the region of the spurs, the proportion of black to white is identical from the center of the disk to its circumference and should produce a uniform gray tone when spun. The white spur should have the effect of making a light gray ring, and the dark spur should make a dark gray ring adjacent to the lighter one, or at least that is what the distribution of intensities predicts. In actuality, what the observer sees are two uniform zones—an inner, light, circular area surrounded by a darker ring. It is as though the dark and light contours somehow carry along or assimilate the uniform gray zones adjacent to them!

To summarize, then, you have seen that contours are enhanced where light and dark meet, and that the contour can sometimes carry over into adjoining regions. There is far less agreement on how the Craik-O'Brien effect works, although a neural mechanism unlike the one for Mach bands is sometimes postulated.

A most extraordinary rotating disk demonstration was published in 1968 by Whitman Richards of the Massachusetts Institute of Technology. When the design is rotated, a ring is formed between the two spike-like spurs of the pattern, against a gray background. What is so intriguing about the display is that viewed at long distances the ring appears lighter than its background, but when viewed at shorter distances the ring looks darker than its background. How is such a thing possible? Whatever the explanation, it must take into account that the distribution of the pattern over space must be related to how light and dark are judged. By "distribution" I mean that at a greater viewing distance the ring occupies a smaller part of the visual field than for nearby viewing.

SPATIAL FREQUENCY THEORY

That the spacing of features on the retina by Richards's disk could be so critical to contrast perception is not all that surprising in light of recent advances in *spatial frequency theory*, a theory that is the focal point of contemporary research on form perception. As Lloyd Kaufman has noted, the problem with even the best feature-detection theory of form perception is that it is too complicated. It requires that there be seemingly endless feature detectors in our brains, at least one for every variation of geometry. There must be line detectors for every orientation of a line, innumerable edge detectors, curve detectors, and on and on. Wouldn't it be better if all of the variations in pattern could be reduced to a central principle? Spatial frequency theory attempts to accomplish just that. It reduces patterns to collections of regular variations in light and dark, much as complex sound waves can be reduced to component, simple sound waves.

The principles of the theory, though unfamiliar to the uninitiated, are not terribly hard to grasp. Suppose that you were presented with a pattern that alternated dark and light regions in a regular fashion. If a light meter were passed across the pattern, it would provide a means for graphing the distribution of intensities at successive spatial locations, just as was done with many of the previous patterns. In this example the regular variations in intensity describe a sine wave when graphed (not to be confused with theoretical depiction of a light wave as a sine wave—the two are unrelated). The printed pattern we have here looks like a series of fuzzy bars and is referred to as a *sine wave grating*, after its intensity graph. If the bars of the grating were few and broad, and the viewer's distance from the pattern fixed, it could be said that the pattern was a relatively *low frequency grating*, particularly when compared with the more numerous bars of a *high frequency grating*. The frequency of a grating is measured by the number of cycles of alternating light and dark per degree of visual angle. This refinement is necessary in order to account for viewing distance.

At this point you may be asking yourself, "What do gratings have to do with form perception?" If gratings of different frequencies are combined, new grating patterns will be formed. Of more immediate importance is the reverse principle, namely, that any existing pattern, no matter how complicated, can be reduced to a combination of simpler spatial frequencies. This theoretical principle is important because it allows for a more parsimonious model of pattern vision than was ever possible before, one which says that form perception can be reduced to a finite number of spatial frequencies.

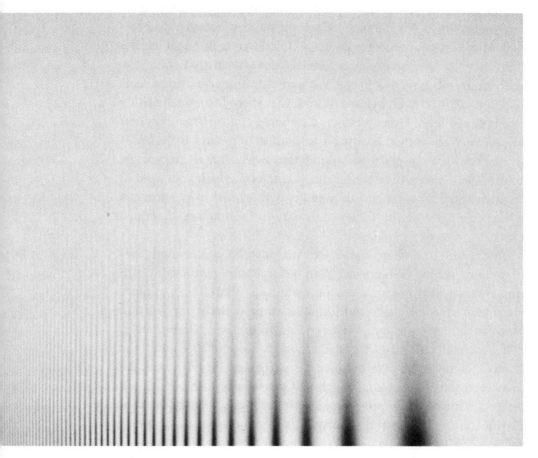

Contrast can be thought of as the ratio of intensities between adjacent regions, and our sensitivity to contrast varies as a function of spatial frequency. This display, created by Fergus Campbell of Cambridge University, varies contrast vertically (with higher contrasts at the bottom of the page and lower contrasts above) and spatial frequency along the horizontal dimension. The sinusoidal distribution of light and dark bars has its highest frequency on the left and lowest frequency on the right. We are most sensitive to contrast at intermediate frequencies, those at the center of the grating. Lower contrast is more difficult to discern at both higher and lower spatial frequencies.

The author wishes to thank Dr. Fergus Campbell, Cambridge University, for his kind permission to reproduce Figure 17.9.

There is convincing experimental evidence that many species of animals, including human beings, respond differently to gratings of different spatial frequencies. Direct measurement of the brain activity of experimental animals has shown that there are individual cells sensitive to specific spatial frequencies, much as it had been shown by Hubel and Wiesel that there are cells sensitive to edges and lines. Human beings also judge contrast as a function of spatial frequency. Sensitivity to contrast appears to be most acute for sinewave gratings around two cycles per degree of visual angle. Sensitivity deteriorates at increasingly higher spatial frequencies. This relationship may be at the heart of phenomena as commonplace as our ability to discern the characters on an eye chart.

To return to Richard's disk, it can now be understood that at a close viewing distance the ring has a much lower spatial frequency than at a greater viewing distance. (It would take more rings packed side-by-side to fill the visual field at the far position than at the near position.) The spatial frequencies determine if sufficient contrast exists for a contour to be perceived, and the contour factor in turn determines the perceived brightness of the area enclosed by contours, much as Cornsweet's illusion demonstrated. Thus spatial frequency theory provides a great deal of clarification for both straightforward and anomalous instances of pattern perception.

Do not be misled into thinking that spatial frequency is the only theory of form perception, or that all of its predictions have been borne out by the data of research. Yet spatial frequency analysis may prove to be the key that will eventually unlock the complexities of form perception.

18. Color Curiosities

If you study visual perception long enough, you will be left with the uneasy feeling that books about perception are traditionally organized around categories that are more than a little bit arbitrary. Form perception cannot be understood independently of contrast effects, motion perception is involved in depth perception, and so on. The classifications that books impose on perceptual phenomena are more for the convenience of author and reader than they are categories imposed by nature.

This being the case, it should not surprise you to learn that the contrast effects mentioned throughout this book have their counterparts in the color dimensions. Not only can one region of intensity affect the brightness of an adjoining region, but the color (hue and saturation) of one area may affect how color is perceived in a neighboring region. Such color interactions were earlier labeled *simultaneous color contrast,* and it is theoretically possible that lateral inhibition and excitation function in color perception much as they do in brightness perception.

The first serious studies of simultaneous color contrast were performed by a French chemist, Michel Eugene Chevreul, whose varied career spanned most of the 1800s. Chevreul, who garnered international recognition for his discoveries in the chemistry of animal fats, was a comfortably established academician before he ever delved into the subject of color. In 1824 Chevreul was ap-

pointed to the position of Director of Dyes for the Royal Manu-
factures at Gobelins, producers of fine tapestries and fabrics. He
was challenged to find out why dye lots varied so much from one
another and in so doing was forced to consider not only dyestuffs,
but also how people see color. He conducted many experiments
on the ways in which one area of color may influence how color
is seen in adjoining regions, a factor that still plagues graphic designers. Chevreul's massive work on the subject, *The Principles of
Harmony and Contrast of Colors and their Applications to the
Arts*, while sometimes grossly incorrect in light of contemporary
knowledge, nonetheless influenced the thinking of scientists like
Helmholtz and scores of artists ranging from Chevreul's contem-
poraries the impressionists to modern artists like Josef Albers.

As suggested in an earlier unit, you may wish to experiment
with color contrast effects with squares of colored paper, much as
Albers instructed his students to do over the years. Start by
examining how two different background colors affect the appearance of smaller patches of equal hue located within them.
Construction paper is acceptable for this purpose, but a wider as-
sortment of colors can be found in graphic arts papers (sold at art
supply stores) or the color samples put out by paint manufac-
turers. The color differences induced by the surrounding colors
are subtle but unmistakable.

Since simultaneous color contrast has been so firmly established,
it is most astonishing to find that a small alteration in the stimulus
pattern can sometimes produce the reverse effect! The action of a
surrounding color is to induce its opposite quality in the surrounded region during simultaneous color contrast, e.g., two identical gray squares take on the complimentary hue and contrasting
brightness of their surrounds. In the case of the *Bezold effect*,
however, a colored region will take on the same color as its surround. Wilhelm von Bezold, a meteorologist, discovered the effect
while practicing his hobby, rug design. The Bezold effect can best
be seen when two areas of identical hue are surrounded by thin
black and white borders respectively. The black-bordered patch
will look slightly darker than the one bordered by white.

There have been a few studies of how the changeover from a

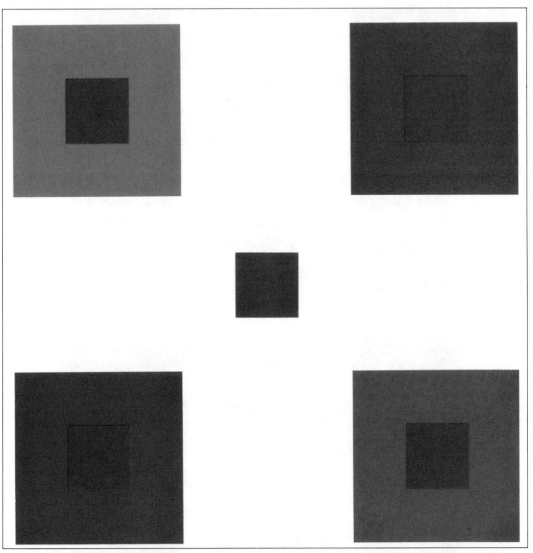

Simultaneous color contrast. The greenish squares are physically identical but they appear to be slightly different from one another when viewed on different backgrounds.

(This page also appears in color following page 20.)

The Bezold effect. Now the green squares do not contrast with the surrounding color but seem to assimilate it; the squares look darker on the
dark background and lighter on the light background.

(This page also appears in color following page 20.)

contrast effect to an assimilation effect takes place, and there is good reason to believe that spatial frequency factors dictate which effect will prevail. In fact, there is more than a passing resemblance between these two effects and the Richards disk of the last unit, a phenomenon in which spatial frequency has also been implicated. It would be interesting to see how the Richards pattern might affect the perception of hue, perhaps by drawing his pattern on a colored disk and observing the rotating disk at different distances.

The color phenomena we've examined so far have a plausible connection to spatial frequency theory, but other color phenomena have more obscure origins. In 1933, for example, M. Luckiesh and F. K. Moss published a pattern of thin, diagonally-oriented black lines on a white field that produce pronounced impressions of color in the observer. Although the perceived colors are not well-saturated, they are most distinct and even take the form of a filigree of hexagonal "cells," which appears to be oriented at right angles to the inducing lines of the pattern. This honeycomb pattern of colors has never been adequately explained, but we do understand that the colors one sees in the Luckiesh-Moss pattern are subjective, since they are not produced by different wavelengths of light but by the visual system of the observer.

Subjective color phenomena of this sort have been reported on for more than a century and a quarter. The Luckiesh-Moss pattern is an unusual example of the genre because the colors are perceived in a stationary pattern while, with few exceptions, other subjective color phenomena have required a moving or flickering stimulus. The basic phenomenon was discovered in 1826 by a French monk, Bénédict Prevost, when he observed the effects of rapidly shuttering a ray of white light with a square of cardboard. The beam of light, observed in a darkened room, could be made to appear violet, green, or yellow by interrupting it with the crude shutter. Subjective colors seen in this way may not be as alien to your own experience as you might at first suppose. Perhaps you have seen brief flashes of color in the whirring of blades of a fan or in the random pattern of "snow" of a black-and-white television set tuned to an empty channel.

The Luckiesh-Moss figure has never been adequately explained. Observers see a faintly colored filigree or honeycomb among the black-and-white diagonals. The orientation of the colored filigree depends on the orientation of the diagonals, as the two variations illustrate.

From M. Luckiesh, *Visual Illusions: Their Causes, Characteristics and Applications.* Copyright © 1965 by Dover Publications, Inc. Used by permission.

There is a great deal of confusion about what to call these flickering phenomena since they have been repeatedly "discovered" by many well-intentioned researchers over the years. Spinning disks were favored by some investigators to create the subjective colors, the most famous of these having been made by the Englishman C. E. Benham at the turn of the century. By incorporating the black and white patterns into a top, Benham invented an extremely popular toy that spawned many imitations. Although the *Benham top* was made famous by its namesake, the technique was actually created by Gustav Fechner some years before Benham. Because the same basic color phenomenon has had such a tortured history, different writers sometimes give it different names. (In an excellent review of the literature, Josef Cohen and Donald A. Gordon went so far as to call it the *Prevost-Fechner-Benham subjective colors!*)

I have reprinted several of the patterns for your entertainment and edification. A pencil poked through the center of each disk will make the patterns into tops. The patterns are simple enough to be modified as you like.

How can we be certain that the colors are indeed produced within the observer rather than by the breakdown of white light into its constituent wavelengths? One way is to take a color photograph of the spinning top. If the color is due to discrete wavelengths, then the color film will be affected as the retina would be. Since the photograph actually looks gray, not colored, it must be concluded that the effect is a function of the visual system itself. The appearance of color persists in the spinning disk even when it is illuminated by a narrowly colored light source or when the pattern is drawn on a colored background.

The most reasonable question to be asked, then, is why do we see color in these patterns? While no one can answer with perfect assurance, one theory of the disks has been supported by considerable experimental evidence. Examine the Benham disk again and consider what happens as a single band on the disk rotates past a fixed point on the observer's retina. As the white section moves past, a relatively high level of light is reflected to the eye, but as the drawn portion comes into view, the light level is reduced. The sequence is repeated many times per second, and assuming that the speed of rotation is fairly constant, a pattern of intensities over time will be created by the disk. Perhaps this "temporal Morse Code" mimics the nervous system's own color code, although no one has ever found the precise location in the visual nervous system that this might occur. The theory, first proposed by Troland in 1922, was more recently elaborated and tested by Leon Festinger, Mark Allyn, and Charles White.

If the outline of the theory is correct, these researchers reasoned, then it is the modulation of light that is the key variable to subjective color, not the fact that a pattern of light moves across the retina. As a way of confirming their suspicions, they had subjects view a stationary white light that was arranged in such a way that it could be made to flicker in much the same temporal order as the reflections from a Benham top. The observer reported seeing color in the stationary light, and by altering the intensity code the experimenters were able to reliably change the reports of specific colors. Theoretically, any device that is capable of rapidly modulating light intensity (like spinning fan blades or a rotating black-and-white disk) could produce the appearance of color. Black-and-white television transmissions qualify, and, according to Festinger, experimental programs have been created that have made black-and-white televisions produce subjective color. These attempts were never commercially successful because the saturation of the colors was weak and variable, and because standard color broadcasting systems were uniformly superior.

The intensity modulation hypothesis is a plausible explanation for the color seen in rotating and flickering achromatic patterns, but is not adequate as an explanation of the color seen in stationary patterns of the Luckiesh-Moss variety. How could a modulation sequence occur in the stationary grid? Do eye movements play a role in the perception of colors in the stationary pattern? As matters now stand, additional research will be needed before we know for sure.

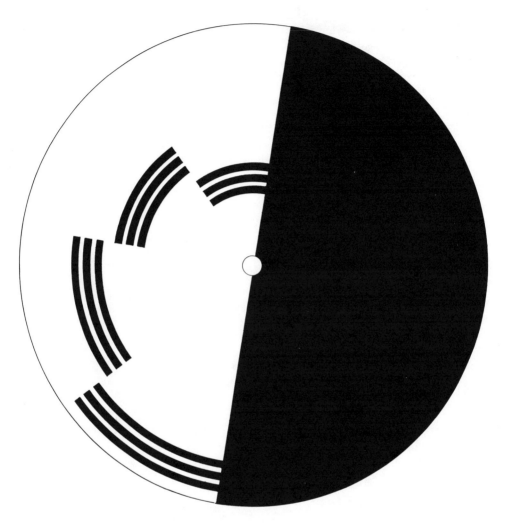

Benham (1894)

Rotating black-and-white "tops" also produce subjective color. Make the disks into tops by mounting on cardboard and poking a sharpened length of dowel or pencil through each one. Each design is accompanied by the name of its originator and the date of its appearance.

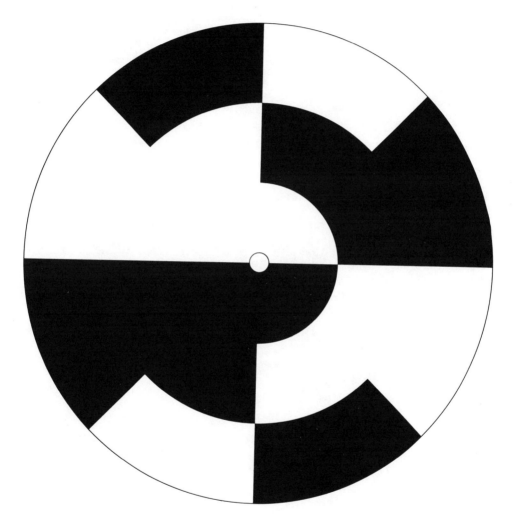

Helmholtz (1856)

Brücke (1864)

Finnegan and Moore (1895)

19. Poggendorff's Illusion

For the last several years I have had an intense interest in the drawn figures called *geometric illusions*. In many ways they sum up everything that I find fascinating and bewildering about visual perception. In this section I will try to explain why it is that these pictures, the sorts of curiosities one normally encounters only in comic books and restaurant placemats, are so central to our understanding of how we see.

Narrow definitions do not work well for the geometric illusions. Not all strange pictures qualify as illusions. The ambiguous pictures of Chapter 15, for example, are not, strictly speaking, illusions. Geometric illusions are drawn figures in which the physical dimensions of size, shape, or direction are consistently misjudged. While many of these pictures portray angular features, not all of them do. One reason that geometric illusions have fascinated researchers is that many of them are so thoroughly uncomplicated in their composition, but so difficult to explain. The *vertical-horizon illusion* consists of only two lines, yet the vertical member always looks longer than its physically equivalent horizontal counterpart. How can such a powerful discrepancy be generated by two straight lines?

The word "illusion" may be misleading since it implies that the discrepancy between the physical measurement of the figure and the way it is judged is an aberration of vision. Quite the opposite is true. The discrepancy is entirely predictable and, to the best of our knowledge, normal for all observers, even experienced ones. When the time comes that researchers understand how geometric illusions work, they will also understand some of the principles of pattern perception.

This last point cannot be stressed strongly enough. In an historical sense, one of the oddest practices in psychology has been the tendency to make grand, all-encompassing theories of behavior without first establishing the underlying principles. This practice is analogous to someone attempting to invent a pocket-sized, transistor radio without first having mastered the principles of electricity. How can we in psychology ever hope to comprehend how human behavior functions without first understanding the most basic processes of perception? Consider the current vogue in psychology to study "person perception." How can we understand how one person perceives another when we cannot even explain to everyone's satisfaction why one line looks longer than another in the configuration of the vertical-horizontal illusion?

The geometric illusions always remind us that when an organism sees, it doesn't just transform the features of its external world into an exact internal "representation." The very nature of the organism is such that it modifies the transformation of external energies into action.

Although hundreds of illusions and their variations exist, it would be much too tedious to catalog them all here. There is no compelling reason to believe that they all have the same origin, or even that a single variable is always the sole cause of a given illusion. The literature of illusion research reveals many factors in the observer or the pattern that contribute to the illusion. This is true even for the simplest geometric illusions.

The vertical-horizontal illusion, like some other famous geometric illusions (the *Müller-Lyer* and *Ponzo illusions*, for example), are illusions of size. That is, two drawn features in one of these pictures appear to be of unequal size when they are measurably equal. The *Orbison illusion* is representative of illusions in which the true shape of a geometric figure appears to be distorted because of its context. The *Poggendorff illusion* is typical of those in

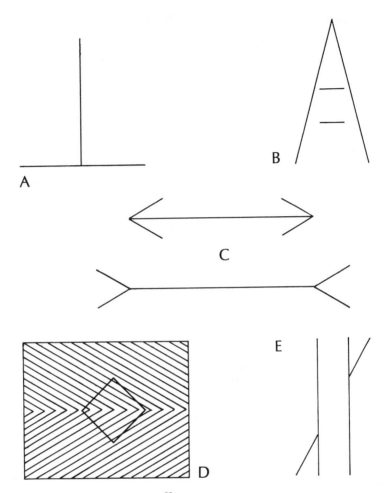

Some well-known geometric illusions:

A. The vertical-horizontal illusion derives its name from the fact that the vertical line looks longer than the physically equivalent horizontal line.

B. Ponzo's illusion. The upper horizontal line looks longer than the lower horizontal but, of course, is not.

C. The Müller-Lyer illusion. Both horizontal segments are of equal extent yet the one bounded by the arrow "tails" looks longer than the one bounded by the arrow "heads."

D. Orbison's illusion. The shape of the square looks distorted when it is superimposed upon a field of intersecting lines.

E. The Poggendorff illusion is an illusion of misalignment. When the two diagonal lines lie along the same straight path, they no longer appear colinear.

which the perceived path of a line deviates from its actual direction. A ruler held against the two diagonal lines of the Poggendorff figure will show that they really do lie along the same straight line.

Without getting too technical, let's examine the Poggendorff illusion more closely. I have selected the Poggendorff as a model of the difficulties researchers have in explaining the geometric illusions, partially because I have a fondness for the pattern for entirely unknown reasons.

The Poggendorff figure, as its title suggests, was contrived by J. C. Poggendorff, probably in the late 1850s. As originally drawn, it consisted of two parallel, vertical lines intersected by two diagonally oriented line segments. It is no exaggeration to say that hundreds of experiments have been performed and scores of theories advanced to explain the perceived nonlinearity of the diagonal line. What I find most frustrating is that as soon as someone comes up with a new explanation of the illusion, someone else invents a variation of the pattern that the new rule cannot explain—at least it seems to work that way.

Many of these explanations try to describe what happens at the intersections of the vertical and diagonal lines. Perhaps, as Ewald Hering suggested in 1861, we tend to exaggerate the effect of acute angles (or underestimate obtuse angles) at the intersections, and so the diagonals appear disjointed. Maybe the illusion has to do with the properties of the eye itself (such as blurring), or is due to lateral inhibition and excitation at the level of the brain. The list of possible causes extends well beyond these few, but it will give you some idea of how even the simplest figure can be attributed to many causes, singly or in combination.

My own involvement with the Poggendorff illusion started some years ago when I was approached by one of my graduate students, Mark Melingonis, who was interested in doing laboratory

It has been theorized that the Poggendorff illusion has its basis in the way we judge obtuse angles. Indeed, the illusion persists when only the obtuse angles are given.

research in perception. He knew of my interest in illusions and suggested that we collaborate on a study of one of them. As an exercise, I asked him to try to create a version of the Poggendorff illusion which, with the fewest number of elements, would still produce the effect of misalignment. At that point neither of us was very familiar with past research, but if we had been, we would have known that many other investigators had adopted much the same strategy.

As long ago as 1893, one of these investigators, Wilhelm Wundt, had proposed that the illusion would be maintained in the complete absence of the vertical lines. That is to say, the two diagonal lines will appear misaligned just by separating them by the distance ordinarily described by the verticals. This parallel-less form of the illusion has been measured by several researchers since Wundt's day, and their results have not been as clear-cut and consistent as might be hoped for. However, it is safe to say that in the absence of the vertical lines, a very small but significant misalignment is seen. Adding the vertical lines, of course, greatly enhances the illusion. So it might be argued that both types of line must be present in order to produce a robust version of the illusion, a circumstance that is in agreement with those theories that concern themselves with the intersections of the vertical and diagonal lines. On the other hand, an illusory contour version of the Poggendorff figure will also create an appreciable illusion, even in the absence of tangible vertical contours.

At that point Melingonis and I wondered whether any lines at all were necessary to produce the effect. We found that several types of reduced figures had been used by other investigators to create Poggendorff's misalignment. In 1971 Daniel Weintraub and

Still another way for the figure to be reduced to its most basic feature is to create a figure in which only the endpoints of the lines remain. In this version, created by Stanley Coren, subjects asked to imagine the missing lines still saw the illusion.

From S. Coren, "Lateral inhibition and geometric illusions," *Quarterly Journal of Psychology*, 1970, 22, 274–278. Used by permission.

David Krantz described an exhaustive series of experiments in which segments of the Poggendorff figure were systematically deleted in order to see how specific parts of the figure affect the magnitude of the illusion. Among other results, they found that the obtuse angles were particularly critical to the illusion, as can be seen in the illustration in which two vertical elements have been removed, leaving only obtuse angles.

At about the same time (1970), Stanley Coren had published a most original impoverished Poggendorff figure, one in which only the end-points of the lines were indicated as a series of dots. His subjects were instructed to imagine the missing contours between the dots and then set the two diagonal segments to be continuous on an adjustable apparatus. Another group of subjects, the control group, did exactly the same task using the traditional Poggendorff drawing. Although the dot version produced a somewhat reduced effect, there was no question but that it did indeed produce the illusion. Coren's experiment sounded a cautionary note for those theories of the illusion that said that it was caused solely by

That vertical lines are not an absolute necessity for the illusion can be seen in an illusory contour version. Even here the diagonals appear misaligned when they are not.

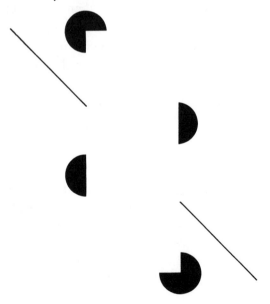

the interaction of lines, whether by lateral inhibition or some other mechanism.

Continuing on this theme, Melingonis and I tried to simplify the illusion further. One way to do this would be to combine the parallel-less version with the dot version, but we suspected that since both were inferior to the traditional figure, their combination would be an extremely fragile effect at best. It then dawned on me that a single dot might work if it were made to move along a diagonal path. (While this insight did not rank with the discovery of penicillin, it was still very important to me in its own way.) Excitedly, I drew a dot on a strip of paper and two parallels on another sheet, then carefully cut an opening in the second piece so that the dot could be pulled behind the verticals along a diagonal path. It was immediately apparent to us that the dot seemed to "jump" between the time it disappeared behind the first vertical and the time it reappeared from beneath the second one! Next, we repeated the crude experiment using a sheet of paper that had no vertical lines, just a pair of slits, and again the diagonally moving dot seemed to veer from a straight path.

The nice thing about working in this area of visual perception is that so much can be understood with tools no more complicated than a pencil, ruler, scissors, and a pad of paper. You ought to try drawing your own variations of the Poggendorff illusion to see how such factors as the angle of the diagonals, the orientation of the figure, and the distance between verticals affect the magnitude of the illusion. You might also investigate how the removal of specific line segments alter the strength of the effect. These variations are not the rigorously controlled experiments that scientists favor, but they can be highly suggestive, just as the moving dot on the strip of paper was to us.

Once we were reasonably certain that we were on to something, we set up an instrumented experiment and compared how people judged a moving dot diagonal and its solid-line counterpart, with and without the vertical lines present. We were quite certain that the moving-dot version would produce an illusion, but were unprepared for the results that we obtained. For one thing, the moving-dot versions of the illusion, with the parallels present or absent, always produced a more powerful effect than for comparable solid-line versions. This was an unusual instance of having the magnitude of the illusion actually increase as the figure was impoverished. Even the version in which there were no verticals, just a dot moving diagonally, was seen to be more misaligned than the customary Poggendorff figure.

What does this mean? I am not exactly sure. One implication of our research is that simple theories of the illusion are not very satisfying. Theories postulating that the effect arises entirely from the interactions of contours are not sophisticated enough to account for our results. Keep in mind that a significant effect occurred when all that a subject saw was a glowing dot that traveled a short distance diagonally, disappeared momentarily, and then reappeared along the second diagonal path. I am afraid that it will take someone much more intelligent that I am to bring order out of the mass of findings that bear upon Poggendorff's illusion.

CUT _____

CUT _____

— CUT — — — — — — — — — — — — — — — — — — —

●

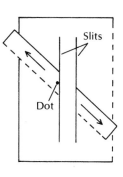

In an attempt to reduce the illusion still further, the author and Mark Melingonis made a version of the illusion in which a single dot traversed a path normally described by the diagonals. We found a very robust version of the illusion, even in the absence of the verticals. To see for yourself, cut out the strip containing the dot and move it through the slits on the page. The dot will appear to shift its position from the time it disappears behind the first vertical line until it reappears from behind the second. Repeat, using slits cut in a blank sheet of paper. The verticals enhance the illusion but their absence does not eliminate it entirely.

Annotated Bibliography

1. A LIGHT AND COLOR PRIMER

Estimates of the number of discriminably different colors were obtained from two sources, F. A. Geldard, *The Human Senses* (2nd ed.), New York: Wiley, 1972, and D. B. Judd and K. L. Kelly, *Color: Universal Language and Dictionary of Names*, National Bureau of Standards Special Publication 440, Washington, D.C.: U.S. Government Printing Office, 1976.

Additional material was obtained from R. M. Boynton, "Color Vision," in J. W. Kling and L. A. Riggs, eds., *Woodworth and Schlosberg's Experimental Psychology*, New York: Holt, Rinehart, and Winston, 1971; and L. A. Riggs, "Vision," in J. W. Kling and L. A. Riggs, eds., *Woodworth and Schlosberg's Experimental Psychology*, New York: Holt, Rinehart, and Winston, 1971.

How color film responds to light at different times of the day can also be seen in J. Hedgcoe, *The Photographer's Handbook*, New York: Knopf, 1978.

2. THE MINIMUM CASE FOR VISION

The vidid description of aviation vertigo was found in C. D. Rousch, "PBM to the Rescue," *Air Progress Aviation Review*, Spring 1979. Metzger's description of the Ganzfeld and Koffka's comments on his work appeared in K. Koffka, *Principles of Gestalt Psychology*, New York: Harcourt, Brace, 1935. The account of blankout appeared in W. Cohen, "Spatial and textural characteristics of the Ganzfeld," *American Journal of Psychology*, 1957, 70, 403–410.

For an excellent review of the Ganzfeld literature, see L. L. Avant, "Vision in the Ganzfeld," *Psychological Bulletin*, 1965, *64*, 246–258. Both the color effects and the table-tennis-ball viewing technique are discussed in J. E. Hochberg, W. Triebel, and G. Seaman, "Color adaptation under conditions of homogeneous visual stimulation (Ganzfeld)," *Journal of Experimental Psychology*, 1951, *41*, 153–159. Interesting physiological effects of the Ganzfeld are discussed in G. Westheimer, "Accommodation Levels in Empty Visual Fields," in E. P. Horne and M. A. Whitcomb, eds., *Vision Research Reports* (Pub. 835), Washington, D.C.: National Academy of Science–National Research Council, 1960.

Sensory deprivation was first discussed in W. H. Bexton, W. Heron, and T. H. Scott, "Effects of decreased variation in the sensory environment," *Canadian Journal of Psychology*, 1954, *8*, 70–76.

3. COMPLETION: PHANTOMS OF THE VISUAL SYSTEM

The subjective Necker cube appeared in D. R. Bradley and H. M. Petry, "Organizational determinants of subjective contour: The subjective Necker cube," *American Journal of Psychology*, 1977, *90*, 253–262. The cube itself was first shown in L. A. Necker, "Observations on some remarkable phenomena seen in Switzerland; and an optical phenomenon which occurs on viewing of a crystal or geometric solid," *Philosophy Magazine*, 1832, *1*, 329–337. The relationship between illusory contour and depth perception appeared in S. Coren, "Subjective contour and apparent depth," *Psychological Review*, 1972, *79*, 359–367. The rotating phantom contour figure was created by R. Sekuler and E. Levinson, "The perception of moving targets," *Scientific American*, 1977, *236*, 60–73.

4. MOTION RECEPTORS

The waterfall illusion was reported in R. Addams, "An account of a peculiar optical phenomenon seen after having looked at a moving body," *Philosophy Magazine*, 1834, *5*, 373.

The eye movement theory of the illusion is discussed in R. S. Woodworth and H. Schlosberg, *Experimental Psychology*, New York: Holt, Rinehart and Winston, 1954.

The motion detector theory is discussed in: E. Levinson and R. Sekuler, "The independence of channels in human vision selective for direction of movement," *Journal of Physiology*, 1975, *250*, 347–366; J. O. Robinson, *The Psychology of Visual Illusions*, London: Hutchinson, 1972 (see pp. 219–233 in particular); R. W. Sekuler and L. Ganz,

"Aftereffect of seen motion with a stabilized retinal image," *Science,* 1963, *139*, 419–420; and R. Sekuler and E. Levinson, "The perception of moving targets," *Scientific American,* 1977, *236*, 60–73.

The early work on feature detectors can be found in D. H. Hubel and T. N. Wiesel, "Receptive fields of single neurones in the cat's striate cortex," *Journal of Physiology,* 1959, *148*, 574–591.

Credit for the spiral drawing techniques is given to B. Gray, *Studio Tips for Artists and Graphic Designers,* New York: Van Nostrand, 1976.

5. THE ILLUSION OF MOVEMENT

Histories of the technology of the cinema appear in C. W. Ceram, *Archaeology of the Cinema,* New York: Harcourt, Brace, and World, 1965, and M. Quigley, *Magic Shadows: The Story of the Origin of Motion Pictures,* New York: Biblo and Tannen, 1969.

The phenakistoscope or fantascope wheel reproduced in this unit is part of a collection held by the Yale Center for British Art, and is described in the exhibit catalog, J. B. Thomas, P. B. Matthews, and D. S. Berman, *The Cottage of Content or, Toys, Games and Amusements of Nineteenth Century England,* New Haven: Yale Center for British Art, 1977. Additional devices of this type can be found in B. Wentz, *Paper Movie Machines,* San Francisco: Troubador Press, 1975.

An excellent discussion of the limitations of persistence of vision as an underlying cause of apparent movement is contained in R. S. Woodworth, *Experimental Psychology,* New York: Holt, 1938.

6. THE WAGON WHEEL EFFECT

The historical material for this chapter was based upon M. Quigley, *Magic Shadows: The Story of the Origin of Motion Pictures,* New York: Biblo and Tannen, 1969 (see in particular the account of Joseph Plateau).

An excellent discussion of the possible relationship between real and apparent motion can be found in L. Kaufman, *Sight and Mind,* New York: Oxford University Press, 1974. That account is based in part upon L. Kaufman, I. Cyrulnick, J. Kaplowitz, G. Melnick, and D. Stof, "The complementarity of apparent and real motion," *Psychologische Forschung,* 1971, *34*, 343–348.

7. MOTION PERCEPTION: LEARNED OR INNATE?

The research described in this section originally appeared in I. Rock, E. S. Tauber, and D. P. Heller, "Perception of stroboscopic movement:

Evidence for its innate basis," *Science*, 1965, *147*, 1050–1052; and E. S. Tauber and S. Koffler, "Optomotor response in human infants to apparent motion: Evidence of innateness," *Science*, 1966, *152*, 382–383.

8. KINETIC ART

An excellent general reference for the topics discussed in this chapter and the next is M. L. Braunstein, *Depth Perception Through Motion*, New York: Academic Press, 1976. Some of the same material appeared in M. L. Braunstein, "The perception of depth through motion," *Psychological Bulletin*, 1962, *59*, 422–433.

For a discussion of the artist's approach to this material, see F. S. Duncan, "Kinetic art: On my psychokinematic objects," *Leonardo*, 1975, *8*, 97–101; and A. Schwarz, *The Complete Works of Marcel Duchamp*, New York: Harry N. Abrams, 1970.

Basic properties of kinetic depth are described in: R. B. Mefferd and B. A. Wieland, "Apparent size-apparent distance relationships in flat stimuli," *Perceptual and Motor Skills*, 1968, *26*, 959–966; C. L. Musatti, "Sui fenomeni stereocinetici," *Archivo Italiano di Psicologia*, 1924, *3*, 105–120; H. Wallach, P. Adams, and A. Weisz, "Circles and derived figures in rotation," *American Journal of Psychology*, 1956, *69*, 48–59; and B. A. Wieland and R. B. Mefferd, Jr., "Perception of depth in rotating objects: 3. Asymmetry and velocity as the determinants of the stereokinetic effect," *Perceptual and Motor Skills*, 1968, *26*, 671–681.

9. MORE KINETIC DEPTH

The kinetic depth effect was described in H. Wallach and D. N. O'Connell, "The kinetic depth effect," *Journal of Experimental Psychology*, 1953, *45*, 205–217. Computer animation investigations of the effect are detailed in M. L. Braunstein, *Depth Perception Through Motion*, New York: Academic Press, 1976, and a related technique that employed cinematic animation is given in B. F. Green, Jr., "Figure coherence in the kinetic depth effect," *Journal of Experimental Psychology*, 1961, *62*, 272–282.

The Ames window illusion appeared in A. Ames, Jr., "Visual perception and the rotating trapezoidal window," *Psychological Monographs*, 1951, *65*, No. 324. The interested reader should also see N. Pastore, "Some remarks on the Ames oscillatory effect," *Psychological Review*, 1952, *59*, 319–323.

The use of the trapezoidal window to investigate pictorial depth perception in children appeared in A. Yonas, W. T. Cleaves, and L. Pettersen, "Development of sensitivity to pictorial depth," *Science*, 1978, *200*, 77–79.

Sinsteden's windmill illusion was described in E. G. Boring, *Sensation and Perception in the History of Experimental Psychology*, New York: Appleton, 1942.

10. LEONARDO'S WINDOW

For an excellent treatise on the relationship between art and science in the Renaissance, see E. Panofsky, *The Codex Huygens and Leonardo da Vinci's Art Theory*, Westport, Conn.: Greenwood Press, 1971. The Alberti quotation appeared in that work. Leonardo's instructions to would-be artists was obtained from J. P. Richter, ed., *The Literary Works of Leonardo da Vinci, Vol. 1* (3rd ed.), London: Phaidon Press, 1970.

Most introductory perception textbooks discuss the static or pictorial depth cues. A more detailed account of the role of texture is contained in J. J. Gibson, *The Perception of the Visual World*, Boston: Houghton Mifflin, 1950.

11. THE SIZE PROBLEM

For an overview of the size-distance problem see Chapter 8 in L. Kaufman, *Perception: The World Transformed*, New York: Oxford University Press, 1979. Judging size under reduction conditions is described in A. H. Holway and E. G. Boring, "Determinants of apparent visual size with distance variant," *American Journal of Psychology*, 1941, *54*, 21–37.

Familiar size and relative size are the subjects of: W. H. Ittelson, "Size as a cue to distance: static localization," *American Journal of Psychology*, 1951, *64*, 54–67; W. H. Ittelson and F. P. Kilpatrick, "Experiments in perception," *Scientific American*, 1951, *185*, 50–55; and I. Rock and S. Ebenholtz, "The relational determination of perceived size," *Psychological Review*, 1959, *66*, 387–401.

The role of instruction in size perception was investigated in A. H. Hastorf, "The influence of suggestion on the relationship between stimulus size and perceived distance," *Journal of Psychology*, 1950, *29*, 195–217.

12. TWO EYES ARE BETTER THAN ONE

Wheatstone's pioneering work appeared in two papers, one published in 1838 and the other in 1852. Only the first is considered here: C. Wheatstone, "On some remarkable, and hitherto unresolved, phenomena of binocular vision," *Philosophical Transactions of the Royal Society of London*, 1838, 371–394.

The history and use of the stereoscope are described in: T. B. Greenslade, Jr., and M. W. Green, "Experiments with stereoscopic images," *Physics Teacher*, 1973, *11*, 215–221; J. Jones, *Wonders of the Stereoscope*, New York: Knopf, 1976; and J. Walker, "The amateur scientist: Moire effects, the kaleidoscope, and other Victorian diversions," *Scientific American*, 1978, *239*, 182–190.

A most scholarly account of stereopsis, with emphasis on random-dot stereograms, is to be found in B. Julesz, *Foundations of Cyclopean Perception*, Chicago: University of Chicago Press, 1971.

13. PULFRICH'S AMAZING PENDULUM

C. Pulfrich described the effect that bears his name in "Die stereoskopie im Dienste der isochromen und heterochromen Photometrie," *Naturwissenschaften*, 1922, *10*, 553–574.

Other references on the subject include: J. T. Enright, "Stereopsis, visual latency, and three-dimensional moving pictures," *American Scientist*, 1970, *58*, 536–545; A. Lit, "The magnitude of the Pulfrich stereophenomenon as a function of binocular differences of intensity at various levels of illumination," *American Journal of Psychology*, 1949, *62*, 159–181; and J. Walker, "The amateur scientist: Visual illusions that can be achieved by putting a dark filter over one eye," *Scientific American*, 1978, *238*, 142–153.

14. THE ANAMORPHIC PUZZLE

For an overview of shape constancy see J. Hochberg, "Perception. I: Color and Shape" in J. W. Kling and L. A. Riggs, eds., *Woodworth and Schlosberg's Experimental Psychology*, New York: Holt, Rinehart, and Winston, 1971.

Conceptual problems of shape constancy are considered in M. H. Pirenne, *Optics, Paintings and Photography*, London: Cambridge University Press, 1975.

A fascinating collection of anamorphic artworks can be viewed in F. Leeman, J. Elffers, and M. Schuyt, *Hidden Images: Games of Perception, Anamorphic Art, Illusion*, New York: Harry N. Abrams, 1975.

15. THE PROBABILITY OF FORM

An excellent general discussion of reversible and improbable figures can be found in Chapter 8 of J. O. Robinson, *The Psychology of Visual Illusions*, London: Hutchinson, 1972.

Impossible figures were described in L. S. Penrose and R. Penrose,

"Impossible objects: A special type of visual illusion," *British Journal of Psychology*, 1958, *49*, 31–33. Gregory's remarks are also of interest in R. L. Gregory, *Eye and Brain*, New York: McGraw-Hill, 1977.

The concept of multistability in form perception appears in A. Attneave, "Triangles as impossible figures," *American Journal of Psychology*, 1968, *81*, 447–453, and F. Attneave, "Multistability in perception," *Scientific American*, 1971, *225*, 62–71.

See T. M. Cowan and R. Pringle, "An investigation of the cues responsible for figure impossibility," *Journal of Experimental Psychology: Human Perception and Performance*, 1978, *4*, 112–120, concerning the ordering of figure impossibility.

Those interested in the work of M. C. Escher should refer to his book *The Graphic Work of M. C. Escher*, New York: Meredith Press, 1967. The quotation was taken from this source. The figure-ground distinction appeared in E. Rubin, *Synoplevede Figurer*, Copenhagen: Gyldendalske, 1915.

16. THE RESTLESS EYE

Thorough reviews of eye movement research as well as speculations concerning the role of eye movements in visual perception can be found in R. A. Monty and J. W. Senders, *Eye Movements and Psychological Processes*, Hillsdale, N.J.: Lawrence Erlbaum Associates, 1976; and A. L. Yarbus, *Eye Movements and Vision*, New York: Plenum, 1967.

Fixation tendency is discussed in L. Kaufman and W. Richards, "Spontaneous fixation tendencies for visual forms," *Perception and Psychophysics*, 1969, *5*, 85–88.

A good source of information about entoptic visual phenomena is M. L. Rubin and G. L. Walls, *Fundamentals of Visual Science*, Springfield, Ill.: Charles C Thomas, 1969.

The camel-through-the-eye-of-a-needle demonstration appeared in T. E. Parks, "Post-retinal visual storage," *American Journal of Psychology*, 1965, *78*, 145–147. The reader should also see I. Rock and F. Halper, "Form perception without a retinal image," *American Journal of Psychology*, 1969, *82*, 425–440.

17. BANDS, GRIDS, AND GRATINGS

For a detailed account of spatial frequency theory, see T. N. Cornsweet, *Visual Perception*, New York: Academic Press, 1970; and Chapter 5 of L. Kaufman, *Perception: The World Transformed*, New York: Oxford University Press, 1979.

An excellent account of Mach's work in vision is contained in
F. Ratliff, "On Mach's contributions to the analysis of sensations," in
R. S. Cohen and R. J. Seeger, eds., *Ernst Mach Physicist and Philoso-
pher*, Dordrecht, Holland: Reidel, 1970.

The pattern created by W. Richards originally appeared in his article
"Illusory reversal of brightness contrast," *Perceptual and Motor Skills*,
1968, *27*, 1169–1170.

18. COLOR CURIOSITIES

M. E. Chevreul's massive nineteenth century work, *The Principles of
Harmony and Contrast of Colors and Their Applications to the Arts*,
has been republished with an introduction and notes by F. Birren, New
York: Reinhold, 1967. The application of simultaneous color contrast
to art is also considered in J. Albers, *Interaction of Color*, New Haven:
Yale University Press, 1975.

Although somewhat dated, a noteworthy review of subjective color
research is to be found in J. Cohen and D. A. Gordon, "The Prevost-
Fechner-Benham subjective colors," *Psychological Bulletin*, 1949, *46*,
97–136. The Morse code theory of the Benham top was first proposed
in L. T. Trolland, "The enigma of color vision," *American Journal of
Physiology*, 1921, *2*, 23–48. The version of the theory cited in this chapter was adapted from L. Festinger, M. R. Allyn, and C. W. White,
"The perception of color with achromatic stimulation," *Vision Research*,
1971, *11*, 591–612. Festinger described the use of subjective color in
black-and-white television transmissions at a meeting, "New Techniques
in Information Display," New York, November 1968.

19. POGGENDORFF'S ILLUSION

The suggestion that we exaggerate acute angles was made by E. Hering, *Beiträge zur Psychologie*, Leipzig: Engelmann, 1861.

The Poggendorff illusion first appeared in an article by E. Burmes-
ter, "Beiträge zur experimentellen Bestimmung geometrisch-optischer
Täuschungen," *Zeitschrift Psychologie*, 1896, *12*, 355–394.

That the Poggendorff illusion might occur in the absence of verticals
was proposed by W. Wundt, *Grundzüge der Physiologischen Psychologie*, Leipzig: Engelmann, 1893. Experimental tests of the parallel-
less Poggendorff appear in: R. H. Day, "The oblique line illusion: The
Poggendorff effect without parallels," *Quarterly Journal of Experimental Psychology*, 1973, *25*, 535–541; R. H. Day and R. G. Dickenson,
"The components of the Poggendorff illusion," *British Journal of Psychology*, 1976, *67*, 537–552; M. B. Goldstein and D. J. Weintraub, "The

parallel-less Poggendorff: Virtual contours put the illusion down but not out," *Perception and Psychophysics,* 1972, *11,* 353–355; and D. J. Weintraub and D. H. Krantz, "The Poggendorff illusion: Amputations, rotations, and other perturbations," *Perception and Psychophysics,* 1971, *10,* 257–264.

The dot version of the illusion appeared in S. Coren, "Lateral inhibition and geometric illusions," *Quarterly Journal of Experimental Psychology,* 1970, *22,* 274–278.

The moving-dot Poggendorff is described in M. B. Fineman and M. P. Melingonis, "The effect of a moving dot transversal on the Poggendorff illusion," *Perception and Psychophysics,* 1977, *21,* 153–156.

Index

A CATALOG OF SELECTED
DOVER BOOKS
IN ALL FIELDS OF INTEREST

A CATALOG OF SELECTED DOVER
BOOKS IN ALL FIELDS OF INTEREST

CONCERNING THE SPIRITUAL IN ART, Wassily Kandinsky. Pioneering work by father of abstract art. Thoughts on color theory, nature of art. Analysis of earlier masters. 12 illustrations. 80pp. of text. 5⅜ × 8½.   23411-8 Pa. $3.95

ANIMALS: 1,419 Copyright-Free Illustrations of Mammals, Birds, Fish, Insects, etc., Jim Harter (ed.). Clear wood engravings present, in extremely lifelike poses, over 1,000 species of animals. One of the most extensive pictorial sourcebooks of its kind. Captions. Index. 284pp. 9 × 12.   23766-4 Pa. $12.95

CELTIC ART: The Methods of Construction, George Bain. Simple geometric techniques for making Celtic interlacements, spirals, Kells-type initials, animals, humans, etc. Over 500 illustrations. 160pp. 9 × 12. (USO)   22923-8 Pa. $9.95

AN ATLAS OF ANATOMY FOR ARTISTS, Fritz Schider. Most thorough reference work on art anatomy in the world. Hundreds of illustrations, including selections from works by Vesalius, Leonardo, Goya, Ingres, Michelangelo, others. 593 illustrations. 192pp. 7⅛ × 10¼.   20241-0 Pa. $9.95

CELTIC HAND STROKE-BY-STROKE (Irish Half-Uncial from "The Book of Kells"): An Arthur Baker Calligraphy Manual, Arthur Baker. Complete guide to creating each letter of the alphabet in distinctive Celtic manner. Covers hand position, strokes, pens, inks, paper, more. Illustrated. 48pp. 8¼ × 11.
                                                                        24336-2 Pa. $3.95

EASY ORIGAMI, John Montroll. Charming collection of 32 projects (hat, cup, pelican, piano, swan, many more) specially designed for the novice origami hobbyist. Clearly illustrated easy-to-follow instructions insure that even beginning papercrafters will achieve successful results. 48pp. 8¼ × 11.   27298-2 Pa. $2.95

THE COMPLETE BOOK OF BIRDHOUSE CONSTRUCTION FOR WOOD-WORKERS, Scott D. Campbell. Detailed instructions, illustrations, tables. Also data on bird habitat and instinct patterns. Bibliography. 3 tables. 63 illustrations in 15 figures. 48pp. 5¼ × 8½.   24407-5 Pa. $1.95

BLOOMINGDALE'S ILLUSTRATED 1886 CATALOG: Fashions, Dry Goods and Housewares, Bloomingdale Brothers. Famed merchants' extremely rare catalog depicting about 1,700 products: clothing, housewares, firearms, dry goods, jewelry, more. Invaluable for dating, identifying vintage items. Also, copyright-free graphics for artists, designers. Co-published with Henry Ford Museum & Greenfield Village. 160pp. 8¼ × 11.   25780-0 Pa. $9.95

HISTORIC COSTUME IN PICTURES, Braun & Schneider. Over 1,450 costumed figures in clearly detailed engravings—from dawn of civilization to end of 19th century. Captions. Many folk costumes. 256pp. 8⅜ × 11¾.   23150-X Pa. $11.95

STICKLEY CRAFTSMAN FURNITURE CATALOGS, Gustav Stickley and L. & J. G. Stickley. Beautiful, functional furniture in two authentic catalogs from 1910. 594 illustrations, including 277 photos, show settles, rockers, armchairs, reclining chairs, bookcases, desks, tables. 183pp. 6½ × 9¼.   23838-5 Pa. $9.95

AMERICAN LOCOMOTIVES IN HISTORIC PHOTOGRAPHS: 1858 to 1949, Ron Ziel (ed.). A rare collection of 126 meticulously detailed official photographs, called "builder portraits," of American locomotives that majestically chronicle the rise of steam locomotive power in America. Introduction. Detailed captions. xi + 129pp. 9 × 12.   27393-8 Pa. $12.95

AMERICA'S LIGHTHOUSES: An Illustrated History, Francis Ross Holland, Jr. Delightfully written, profusely illustrated fact-filled survey of over 200 American lighthouses since 1716. History, anecdotes, technological advances, more. 240pp. 8 × 10¾.   25576-X Pa. $11.95

TOWARDS A NEW ARCHITECTURE, Le Corbusier. Pioneering manifesto by founder of "International School." Technical and aesthetic theories, views of industry, economics, relation of form to function, "mass-production split" and much more. Profusely illustrated. 320pp. 6⅛ × 9¼. (USO)   25023-7 Pa. $9.95

HOW THE OTHER HALF LIVES, Jacob Riis. Famous journalistic record, exposing poverty and degradation of New York slums around 1900, by major social reformer. 100 striking and influential photographs. 233pp. 10 × 7⅞.
22012-5 Pa $10.95

FRUIT KEY AND TWIG KEY TO TREES AND SHRUBS, William M. Harlow. One of the handiest and most widely used identification aids. Fruit key covers 120 deciduous and evergreen species; twig key 160 deciduous species. Easily used. Over 300 photographs. 126pp. 5⅜ × 8½.   20511-8 Pa. $3.95

COMMON BIRD SONGS, Dr. Donald J. Borror. Songs of 60 most common U.S. birds: robins, sparrows, cardinals, bluejays, finches, more—arranged in order of increasing complexity. Up to 9 variations of songs of each species.
Cassette and manual 99911-4 $8.95

ORCHIDS AS HOUSE PLANTS, Rebecca Tyson Northen. Grow cattleyas and many other kinds of orchids—in a window, in a case, or under artificial light. 63 illustrations. 148pp. 5⅜ × 8½.   23261-1 Pa. $4.95

MONSTER MAZES, Dave Phillips. Masterful mazes at four levels of difficulty. Avoid deadly perils and evil creatures to find magical treasures. Solutions for all 32 exciting illustrated puzzles. 48pp. 8¼ × 11.   26005-4 Pa. $2.95

MOZART'S DON GIOVANNI (DOVER OPERA LIBRETTO SERIES), Wolfgang Amadeus Mozart. Introduced and translated by Ellen H. Bleiler. Standard Italian libretto, with complete English translation. Convenient and thoroughly portable—an ideal companion for reading along with a recording or the performance itself. Introduction. List of characters. Plot summary. 121pp. 5¼ × 8½.
24944-1 Pa. $2.95

TECHNICAL MANUAL AND DICTIONARY OF CLASSICAL BALLET, Gail Grant. Defines, explains, comments on steps, movements, poses and concepts. 15-page pictorial section. Basic book for student, viewer. 127pp. 5⅜ × 8½.
21843-0 Pa. $4.95

CATALOG OF DOVER BOOKS

BRASS INSTRUMENTS: Their History and Development, Anthony Baines. Authoritative, updated survey of the evolution of trumpets, trombones, bugles, cornets, French horns, tubas and other brass wind instruments. Over 140 illustrations and 48 music examples. Corrected and updated by author. New preface. Bibliography. 320pp. 5⅜ × 8½.   27574-4 Pa. $9.95

HOLLYWOOD GLAMOR PORTRAITS, John Kobal (ed.). 145 photos from 1926–49. Harlow, Gable, Bogart, Bacall; 94 stars in all. Full background on photographers, technical aspects. 160pp. 8⅜ × 11¼.   23352-9 Pa. $11.95

MAX AND MORITZ, Wilhelm Busch. Great humor classic in both German and English. Also 10 other works: "Cat and Mouse," "Plisch and Plumm," etc. 216pp. 5⅜ × 8½.   20181-3 Pa. $5.95

THE RAVEN AND OTHER FAVORITE POEMS, Edgar Allan Poe. Over 40 of the author's most memorable poems: "The Bells," "Ulalume," "Israfel," "To Helen," "The Conqueror Worm," "Eldorado," "Annabel Lee," many more. Alphabetic lists of titles and first lines. 64pp. 5³⁄₁₆ × 8¼.   26685-0 Pa. $1.00

SEVEN SCIENCE FICTION NOVELS, H. G. Wells. The standard collection of the great novels. Complete, unabridged. First Men in the Moon, Island of Dr. Moreau, War of the Worlds, Food of the Gods, Invisible Man, Time Machine, In the Days of the Comet. Total of 1,015pp. 5⅜ × 8½. (USO)   20264-X Clothbd. $29.95

AMULETS AND SUPERSTITIONS, E. A. Wallis Budge. Comprehensive discourse on origin, powers of amulets in many ancient cultures: Arab, Persian, Babylonian, Assyrian, Egyptian, Gnostic, Hebrew, Phoenician, Syriac, etc. Covers cross, swastika, crucifix, seals, rings, stones, etc. 584pp. 5⅜ × 8½. 23573-4 Pa. $12.95

RUSSIAN STORIES/PYCCKNE PACCKA3bl: A Dual-Language Book, edited by Gleb Struve. Twelve tales by such masters as Chekhov, Tolstoy, Dostoevsky, Pushkin, others. Excellent word-for-word English translations on facing pages, plus teaching and study aids, Russian/English vocabulary, biographical/critical introductions, more. 416pp. 5⅜ × 8½.   26244-8 Pa. $8.95

PHILADELPHIA THEN AND NOW: 60 Sites Photographed in the Past and Present, Kenneth Finkel and Susan Oyama. Rare photographs of City Hall, Logan Square, Independence Hall, Betsy Ross House, other landmarks juxtaposed with contemporary views. Captures changing face of historic city. Introduction. Captions. 128pp. 8¼ × 11.   25790-8 Pa. $9.95

AIA ARCHITECTURAL GUIDE TO NASSAU AND SUFFOLK COUNTIES, LONG ISLAND, The American Institute of Architects, Long Island Chapter, and the Society for the Preservation of Long Island Antiquities. Comprehensive, well-researched and generously illustrated volume brings to life over three centuries of Long Island's great architectural heritage. More than 240 photographs with authoritative, extensively detailed captions. 176pp. 8¼ × 11.   26946-9 Pa. $14.95

NORTH AMERICAN INDIAN LIFE: Customs and Traditions of 23 Tribes, Elsie Clews Parsons (ed.). 27 fictionalized essays by noted anthropologists examine religion, customs, government, additional facets of life among the Winnebago, Crow, Zuni, Eskimo, other tribes. 480pp. 6⅛ × 9¼.   27377-6 Pa. $10.95

FRANK LLOYD WRIGHT'S HOLLYHOCK HOUSE, Donald Hoffmann. Lavishly illustrated, carefully documented study of one of Wright's most controversial residential designs. Over 120 photographs, floor plans, elevations, etc. Detailed perceptive text by noted Wright scholar. Index. 128pp. 9¼ × 10¾.
27133-1 Pa. $11.95

THE MALE AND FEMALE FIGURE IN MOTION: 60 Classic Photographic Sequences, Eadweard Muybridge. 60 true-action photographs of men and women walking, running, climbing, bending, turning, etc., reproduced from rare 19th-century masterpiece. vi + 121pp. 9 × 12.
24745-7 Pa. $10.95

1001 QUESTIONS ANSWERED ABOUT THE SEASHORE, N. J. Berrill and Jacquelyn Berrill. Queries answered about dolphins, sea snails, sponges, starfish, fishes, shore birds, many others. Covers appearance, breeding, growth, feeding, much more. 305pp. 5¼ × 8¼.
23366-9 Pa. $7.95

GUIDE TO OWL WATCHING IN NORTH AMERICA, Donald S. Heintzelman. Superb guide offers complete data and descriptions of 19 species: barn owl, screech owl, snowy owl, many more. Expert coverage of owl-watching equipment, conservation, migrations and invasions, etc. Guide to observing sites. 84 illustrations. xiii + 193pp. 5⅜ × 8½.
27344-X Pa. $8.95

MEDICINAL AND OTHER USES OF NORTH AMERICAN PLANTS: A Historical Survey with Special Reference to the Eastern Indian Tribes, Charlotte Erichsen-Brown. Chronological historical citations document 500 years of usage of plants, trees, shrubs native to eastern Canada, northeastern U.S. Also complete identifying information. 343 illustrations. 544pp. 6½ × 9¼.
25951-X Pa. $12.95

STORYBOOK MAZES, Dave Phillips. 23 stories and mazes on two-page spreads: Wizard of Oz, Treasure Island, Robin Hood, etc. Solutions. 64pp. 8¼ × 11.
23628-5 Pa. $2.95

NEGRO FOLK MUSIC, U.S.A., Harold Courlander. Noted folklorist's scholarly yet readable analysis of rich and varied musical tradition. Includes authentic versions of over 40 folk songs. Valuable bibliography and discography. xi + 324pp. 5⅜ × 8½.
27350-4 Pa. $7.95

MOVIE-STAR PORTRAITS OF THE FORTIES, John Kobal (ed.). 163 glamor, studio photos of 106 stars of the 1940s: Rita Hayworth, Ava Gardner, Marlon Brando, Clark Gable, many more. 176pp. 8⅜ × 11¼.
23546-7 Pa. $11.95

BENCHLEY LOST AND FOUND, Robert Benchley. Finest humor from early 30s, about pet peeves, child psychologists, post office and others. Mostly unavailable elsewhere. 73 illustrations by Peter Arno and others. 183pp. 5⅜ × 8½.
22410-4 Pa. $5.95

YEKL and THE IMPORTED BRIDEGROOM AND OTHER STORIES OF YIDDISH NEW YORK, Abraham Cahan. Film Hester Street based on Yekl (1896). Novel, other stories among first about Jewish immigrants on N.Y.'s East Side. 240pp. 5⅜ × 8½.
22427-9 Pa. $6.95

SELECTED POEMS, Walt Whitman. Generous sampling from *Leaves of Grass.* Twenty-four poems include "I Hear America Singing," "Song of the Open Road," "I Sing the Body Electric," "When Lilacs Last in the Dooryard Bloom'd," "O Captain! My Captain!"—all reprinted from an authoritative edition. Lists of titles and first lines. 128pp. 5³⁄₁₆ × 8¼.
26878-0 Pa. $1.00

THE BEST TALES OF HOFFMANN, E. T. A. Hoffmann. 10 of Hoffmann's most important stories: "Nutcracker and the King of Mice," "The Golden Flowerpot," etc. 458pp. 5⅜ × 8½. 21793-0 Pa. $8.95

FROM FETISH TO GOD IN ANCIENT EGYPT, E. A. Wallis Budge. Rich detailed survey of Egyptian conception of "God" and gods, magic, cult of animals, Osiris, more. Also, superb English translations of hymns and legends. 240 illustrations. 545pp. 5⅜ × 8½. 25803-3 Pa. $11.95

FRENCH STORIES/CONTES FRANÇAIS: A Dual-Language Book, Wallace Fowlie. Ten stories by French masters, Voltaire to Camus: "Micromegas" by Voltaire; "The Atheist's Mass" by Balzac; "Minuet" by de Maupassant; "The Guest" by Camus, six more. Excellent English translations on facing pages. Also French-English vocabulary list, exercises, more. 352pp. 5⅜ × 8½. 26443-2 Pa. $8.95

CHICAGO AT THE TURN OF THE CENTURY IN PHOTOGRAPHS: 122 Historic Views from the Collections of the Chicago Historical Society, Larry A. Viskochil. Rare large-format prints offer detailed views of City Hall, State Street, the Loop, Hull House, Union Station, many other landmarks, circa 1904–1913. Introduction. Captions. Maps. 144pp. 9⅜ × 12¼. 24656-6 Pa. $12.95

OLD BROOKLYN IN EARLY PHOTOGRAPHS, 1865–1929, William Lee Younger. Luna Park, Gravesend race track, construction of Grand Army Plaza, moving of Hotel Brighton, etc. 157 previously unpublished photographs. 165pp. 8⅞ × 11¾. 23587-4 Pa. $13.95

THE MYTHS OF THE NORTH AMERICAN INDIANS, Lewis Spence. Rich anthology of the myths and legends of the Algonquins, Iroquois, Pawnees and Sioux, prefaced by an extensive historical and ethnological commentary. 36 illustrations. 480pp. 5⅜ × 8½. 25967-6 Pa. $8.95

AN ENCYCLOPEDIA OF BATTLES: Accounts of Over 1,560 Battles from 1479 B.C. to the Present, David Eggenberger. Essential details of every major battle in recorded history from the first battle of Megiddo in 1479 B.C. to Grenada in 1984. List of Battle Maps. New Appendix covering the years 1967–1984. Index. 99 illustrations. 544pp. 6½ × 9¼. 24913-1 Pa. $14.95

SAILING ALONE AROUND THE WORLD, Captain Joshua Slocum. First man to sail around the world, alone, in small boat. One of great feats of seamanship told in delightful manner. 67 illustrations. 294pp. 5⅜ × 8½. 20326-3 Pa. $5.95

ANARCHISM AND OTHER ESSAYS, Emma Goldman. Powerful, penetrating, prophetic essays on direct action, role of minorities, prison reform, puritan hypocrisy, violence, etc. 271pp. 5⅜ × 8½. 22484-8 Pa. $5.95

MYTHS OF THE HINDUS AND BUDDHISTS, Ananda K. Coomaraswamy and Sister Nivedita. Great stories of the epics; deeds of Krishna, Shiva, taken from puranas, Vedas, folk tales; etc. 32 illustrations. 400pp. 5⅜ × 8½. 21759-0 Pa. $9.95

BEYOND PSYCHOLOGY, Otto Rank. Fear of death, desire of immortality, nature of sexuality, social organization, creativity, according to Rankian system. 291pp. 5⅜ × 8½. 20485-5 Pa. $8.95

A THEOLOGICO-POLITICAL TREATISE, Benedict Spinoza. Also contains unfinished Political Treatise. Great classic on religious liberty, theory of government on common consent. R. Elwes translation. Total of 421pp. 5⅜ × 8½. 20249-6 Pa. $8.95

MY BONDAGE AND MY FREEDOM, Frederick Douglass. Born a slave, Douglass became outspoken force in antislavery movement. The best of Douglass' autobiographies. Graphic description of slave life. 464pp. 5⅜ × 8½.   22457-0 Pa. $8.95

FOLLOWING THE EQUATOR: A Journey Around the World, Mark Twain. Fascinating humorous account of 1897 voyage to Hawaii, Australia, India, New Zealand, etc. Ironic, bemused reports on peoples, customs, climate, flora and fauna, politics, much more. 197 illustrations. 720pp. 5⅜ × 8½.   26113-1 Pa. $15.95

THE PEOPLE CALLED SHAKERS, Edward D. Andrews. Definitive study of Shakers: origins, beliefs, practices, dances, social organization, furniture and crafts, etc. 33 illustrations. 351pp. 5⅜ × 8½.   21081-2 Pa. $8.95

THE MYTHS OF GREECE AND ROME, H. A. Guerber. A classic of mythology, generously illustrated, long prized for its simple, graphic, accurate retelling of the principal myths of Greece and Rome, and for its commentary on their origins and significance. With 64 illustrations by Michelangelo, Raphael, Titian, Rubens, Canova, Bernini and others. 480pp. 5⅜ × 8½.   27584-1 Pa. $9.95

PSYCHOLOGY OF MUSIC, Carl E. Seashore. Classic work discusses music as a medium from psychological viewpoint. Clear treatment of physical acoustics, auditory apparatus, sound perception, development of musical skills, nature of musical feeling, host of other topics. 88 figures. 408pp. 5⅜ × 8½. 21851-1 Pa. $9.95

THE PHILOSOPHY OF HISTORY, Georg W. Hegel. Great classic of Western thought develops concept that history is not chance but rational process, the evolution of freedom. 457pp. 5⅜ × 8½.   20112-0 Pa. $9.95

THE BOOK OF TEA, Kakuzo Okakura. Minor classic of the Orient: entertaining, charming explanation, interpretation of traditional Japanese culture in terms of tea ceremony. 94pp. 5⅜ × 8½.   20070-1 Pa. $3.95

LIFE IN ANCIENT EGYPT, Adolf Erman. Fullest, most thorough, detailed older account with much not in more recent books, domestic life, religion, magic, medicine, commerce, much more. Many illustrations reproduce tomb paintings, carvings, hieroglyphs, etc. 597pp. 5⅜ × 8½.   22632-8 Pa. $10.95

SUNDIALS, Their Theory and Construction, Albert Waugh. Far and away the best, most thorough coverage of ideas, mathematics concerned, types, construction, adjusting anywhere. Simple, nontechnical treatment allows even children to build several of these dials. Over 100 illustrations. 230pp. 5⅜ × 8½.   22947-5 Pa. $7.95

DYNAMICS OF FLUIDS IN POROUS MEDIA, Jacob Bear. For advanced students of ground water hydrology, soil mechanics and physics, drainage and irrigation engineering, and more. 335 illustrations. Exercises, with answers. 784pp. 6⅛ × 9¼.   65675-6 Pa. $19.95

SONGS OF EXPERIENCE: Facsimile Reproduction with 26 Plates in Full Color, William Blake. 26 full-color plates from a rare 1826 edition. Includes "The Tyger," "London," "Holy Thursday," and other poems. Printed text of poems. 48pp. 5¼ × 7.   24636-1 Pa. $4.95

OLD-TIME VIGNETTES IN FULL COLOR, Carol Belanger Grafton (ed.). Over 390 charming, often sentimental illustrations, selected from archives of Victorian graphics—pretty women posing, children playing, food, flowers, kittens and puppies, smiling cherubs, birds and butterflies, much more. All copyright-free. 48pp. 9¼ × 12¼.   27269-9 Pa. $5.95

PERSPECTIVE FOR ARTISTS, Rex Vicat Cole. Depth, perspective of sky and sea, shadows, much more, not usually covered. 391 diagrams, 81 reproductions of drawings and paintings. 279pp. 5⅜ × 8½. 22487-2 Pa. $6.95

DRAWING THE LIVING FIGURE, Joseph Sheppard. Innovative approach to artistic anatomy focuses on specifics of surface anatomy, rather than muscles and bones. Over 170 drawings of live models in front, back and side views, and in widely varying poses. Accompanying diagrams. 177 illustrations. Introduction. Index. 144pp. 8⅜ × 11¼. 26723-7 Pa. $8.95

GOTHIC AND OLD ENGLISH ALPHABETS: 100 Complete Fonts, Dan X. Solo. Add power, elegance to posters, signs, other graphics with 100 stunning copyright-free alphabets: Blackstone, Dolbey, Germania, 97 more—including many lower-case, numerals, punctuation marks. 104pp. 8⅛ × 11. 24695-7 Pa. $8.95

HOW TO DO BEADWORK, Mary White. Fundamental book on craft from simple projects to five-bead chains and woven works. 106 illustrations. 142pp. 5⅜ × 8. 20697-1 Pa. $4.95

THE BOOK OF WOOD CARVING, Charles Marshall Sayers. Finest book for beginners discusses fundamentals and offers 34 designs. "Absolutely first rate . . . well thought out and well executed."—E. J. Tangerman. 118pp. 7¾ × 10⅝. 23654-4 Pa. $5.95

ILLUSTRATED CATALOG OF CIVIL WAR MILITARY GOODS: Union Army Weapons, Insignia, Uniform Accessories, and Other Equipment, Schuyler, Hartley, and Graham. Rare, profusely illustrated 1846 catalog includes Union Army uniform and dress regulations, arms and ammunition, coats, insignia, flags, swords, rifles, etc. 226 illustrations. 160pp. 9 × 12. 24939-5 Pa. $10.95

WOMEN'S FASHIONS OF THE EARLY 1900s: An Unabridged Republication of "New York Fashions, 1909," National Cloak & Suit Co. Rare catalog of mail-order fashions documents women's and children's clothing styles shortly after the turn of the century. Captions offer full descriptions, prices. Invaluable resource for fashion, costume historians. Approximately 725 illustrations. 128pp. 8⅜ × 11¼. 27276-1 Pa. $11.95

THE 1912 AND 1915 GUSTAV STICKLEY FURNITURE CATALOGS, Gustav Stickley. With over 200 detailed illustrations and descriptions, these two catalogs are essential reading and reference materials and identification guides for Stickley furniture. Captions cite materials, dimensions and prices. 112pp. 6½ × 9¼. 26676-1 Pa. $9.95

EARLY AMERICAN LOCOMOTIVES, John H. White, Jr. Finest locomotive engravings from early 19th century: historical (1804–74), main-line (after 1870), special, foreign, etc. 147 plates. 142pp. 11⅜ × 8¼. 22772-3 Pa. $10.95

THE TALL SHIPS OF TODAY IN PHOTOGRAPHS, Frank O. Braynard. Lavishly illustrated tribute to nearly 100 majestic contemporary sailing vessels: Amerigo Vespucci, Clearwater, Constitution, Eagle, Mayflower, Sea Cloud, Victory, many more. Authoritative captions provide statistics, background on each ship. 190 black-and-white photographs and illustrations. Introduction. 128pp. 8⅞ × 11¾. 27163-3 Pa. $13.95

EARLY NINETEENTH-CENTURY CRAFTS AND TRADES, Peter Stockham (ed.). Extremely rare 1807 volume describes to youngsters the crafts and trades of the day: brickmaker, weaver, dressmaker, bookbinder, ropemaker, saddler, many more. Quaint prose, charming illustrations for each craft. 20 black-and-white line illustrations. 192pp. 4⅝ × 6.   27293-1 Pa. $4.95

VICTORIAN FASHIONS AND COSTUMES FROM HARPER'S BAZAR, 1867–1898, Stella Blum (ed.). Day costumes, evening wear, sports clothes, shoes, hats, other accessories in over 1,000 detailed engravings. 320pp. 9⅜ × 12¼.
22990-4 Pa. $13.95

GUSTAV STICKLEY, THE CRAFTSMAN, Mary Ann Smith. Superb study surveys broad scope of Stickley's achievement, especially in architecture. Design philosophy, rise and fall of the Craftsman empire, descriptions and floor plans for many Craftsman houses, more. 86 black-and-white halftones. 31 line illustrations. Introduction. 208pp. 6½ × 9¼.   27210-9 Pa. $9.95

THE LONG ISLAND RAIL ROAD IN EARLY PHOTOGRAPHS, Ron Ziel. Over 220 rare photos, informative text document origin (1844) and development of rail service on Long Island. Vintage views of early trains, locomotives, stations, passengers, crews, much more. Captions. 8⅞ × 11¾.   26301-0 Pa. $13.95

THE BOOK OF OLD SHIPS: From Egyptian Galleys to Clipper Ships, Henry B. Culver. Superb, authoritative history of sailing vessels, with 80 magnificent line illustrations. Galley, bark, caravel, longship, whaler, many more. Detailed, informative text on each vessel by noted naval historian. Introduction. 256pp. 5⅜ × 8½.   27332-6 Pa. $6.95

TEN BOOKS ON ARCHITECTURE, Vitruvius. The most important book ever written on architecture. Early Roman aesthetics, technology, classical orders, site selection, all other aspects. Morgan translation. 331pp. 5⅜ × 8½. 20645-9 Pa. $8.95

THE HUMAN FIGURE IN MOTION, Eadweard Muybridge. More than 4,500 stopped-action photos, in action series, showing undraped men, women, children jumping, lying down, throwing, sitting, wrestling, carrying, etc. 390pp. 7⅞ × 10⅝.
20204-6 Clothbd. $24.95

TREES OF THE EASTERN AND CENTRAL UNITED STATES AND CANADA, William M. Harlow. Best one-volume guide to 140 trees. Full descriptions, woodlore, range, etc. Over 600 illustrations. Handy size. 288pp. 4½ × 6⅜.
20395-6 Pa. $5.95

SONGS OF WESTERN BIRDS, Dr. Donald J. Borror. Complete song and call repertoire of 60 western species, including flycatchers, juncoes, cactus wrens, many more—includes fully illustrated booklet.   Cassette and manual 99913-0 $8.95

GROWING AND USING HERBS AND SPICES, Milo Miloradovich. Versatile handbook provides all the information needed for cultivation and use of all the herbs and spices available in North America. 4 illustrations. Index. Glossary. 236pp. 5⅜ × 8½.   25058-X Pa. $6.95

BIG BOOK OF MAZES AND LABYRINTHS, Walter Shepherd. 50 mazes and labyrinths in all—classical, solid, ripple, and more—in one great volume. Perfect inexpensive puzzler for clever youngsters. Full solutions. 112pp. 8⅛ × 11.
22951-3 Pa. $4.95

PIANO TUNING, J. Cree Fischer. Clearest, best book for beginner, amateur. Simple repairs, raising dropped notes, tuning by easy method of flattened fifths. No previous skills needed. 4 illustrations. 201pp. 5⅜ × 8½. 23267-0 Pa. $5.95

A SOURCE BOOK IN THEATRICAL HISTORY, A. M. Nagler. Contemporary observers on acting, directing, make-up, costuming, stage props, machinery, scene design, from Ancient Greece to Chekhov. 611pp. 5⅜ × 8½. 20515-0 Pa. $11.95

THE COMPLETE NONSENSE OF EDWARD LEAR, Edward Lear. All nonsense limericks, zany alphabets, Owl and Pussycat, songs, nonsense botany, etc., illustrated by Lear. Total of 320pp. 5⅜ × 8½. (USO) 20167-8 Pa. $6.95

VICTORIAN PARLOUR POETRY: An Annotated Anthology, Michael R. Turner. 117 gems by Longfellow, Tennyson, Browning, many lesser-known poets. "The Village Blacksmith," "Curfew Must Not Ring Tonight," "Only a Baby Small," dozens more, often difficult to find elsewhere. Index of poets, titles, first lines. xxiii + 325pp. 5⅜ × 8¼. 27044-0 Pa. $8.95

DUBLINERS, James Joyce. Fifteen stories offer vivid, tightly focused observations of the lives of Dublin's poorer classes. At least one, "The Dead," is considered a masterpiece. Reprinted complete and unabridged from standard edition. 160pp. 5³⁄₁₆ × 8¼. 26870-5 Pa. $1.00

THE HAUNTED MONASTERY and THE CHINESE MAZE MURDERS, Robert van Gulik. Two full novels by van Gulik, set in 7th-century China, continue adventures of Judge Dee and his companions. An evil Taoist monastery, seemingly supernatural events; overgrown topiary maze hides strange crimes. 27 illustrations. 328pp. 5⅜ × 8½. 23502-5 Pa. $7.95

THE BOOK OF THE SACRED MAGIC OF ABRAMELIN THE MAGE, translated by S. MacGregor Mathers. Medieval manuscript of ceremonial magic. Basic document in Aleister Crowley, Golden Dawn groups. 268pp. 5⅜ × 8½. 23211-5 Pa. $8.95

NEW RUSSIAN-ENGLISH AND ENGLISH-RUSSIAN DICTIONARY, M. A. O'Brien. This is a remarkably handy Russian dictionary, containing a surprising amount of information, including over 70,000 entries. 366pp. 4½ × 6⅛. 20208-9 Pa. $9.95

HISTORIC HOMES OF THE AMERICAN PRESIDENTS, Second, Revised Edition, Irvin Haas. A traveler's guide to American Presidential homes, most open to the public, depicting and describing homes occupied by every American President from George Washington to George Bush. With visiting hours, admission charges, travel routes. 175 photographs. Index. 160pp. 8¼ × 11. 26751-2 Pa. $10.95

NEW YORK IN THE FORTIES, Andreas Feininger. 162 brilliant photographs by the well-known photographer, formerly with *Life* magazine. Commuters, shoppers, Times Square at night, much else from city at its peak. Captions by John von Hartz. 181pp. 9¼ × 10¾. 23585-8 Pa. $12.95

INDIAN SIGN LANGUAGE, William Tomkins. Over 525 signs developed by Sioux and other tribes. Written instructions and diagrams. Also 290 pictographs. 111pp. 6⅛ × 9¼. 22029-X Pa. $3.50

ANATOMY: A Complete Guide for Artists, Joseph Sheppard. A master of figure drawing shows artists how to render human anatomy convincingly. Over 460 illustrations. 224pp. 8⅜ × 11¼. 27279-6 Pa. $10.95

MEDIEVAL CALLIGRAPHY: Its History and Technique, Marc Drogin. Spirited history, comprehensive instruction manual covers 13 styles (ca. 4th century thru 15th). Excellent photographs; directions for duplicating medieval techniques with modern tools. 224pp. 8⅜ × 11¼. 26142-5 Pa. $11.95

DRIED FLOWERS: How to Prepare Them, Sarah Whitlock and Martha Rankin. Complete instructions on how to use silica gel, meal and borax, perlite aggregate, sand and borax, glycerine and water to create attractive permanent flower arrangements. 12 illustrations. 32pp. 5⅜ × 8½. 21802-3 Pa. $1.00

EASY-TO-MAKE BIRD FEEDERS FOR WOODWORKERS, Scott D. Campbell. Detailed, simple-to-use guide for designing, constructing, caring for and using feeders. Text, illustrations for 12 classic and contemporary designs. 96pp. 5⅜ × 8½. 25847-5 Pa. $2.95

OLD-TIME CRAFTS AND TRADES, Peter Stockham. An 1807 book created to teach children about crafts and trades open to them as future careers. It describes in detailed, nontechnical terms 24 different occupations, among them coachmaker, gardener, hairdresser, lacemaker, shoemaker, wheelwright, copper-plate printer, milliner, trunkmaker, merchant and brewer. Finely detailed engravings illustrate each occupation. 192pp. 4⅝ × 6. 27398-9 Pa. $4.95

THE HISTORY OF UNDERCLOTHES, C. Willett Cunnington and Phyllis Cunnington. Fascinating, well-documented survey covering six centuries of English undergarments, enhanced with over 100 illustrations: 12th-century laced-up bodice, footed long drawers (1795), 19th-century bustles, 19th-century corsets for men, Victorian "bust improvers," much more. 272pp. 5⅜ × 8¼. 27124-2 Pa. $9.95

ARTS AND CRAFTS FURNITURE: The Complete Brooks Catalog of 1912, Brooks Manufacturing Co. Photos and detailed descriptions of more than 150 now very collectible furniture designs from the Arts and Crafts movement depict davenports, settees, buffets, desks, tables, chairs, bedsteads, dressers and more, all built of solid, quarter-sawed oak. Invaluable for students and enthusiasts of antiques, Americana and the decorative arts. 80pp. 6½ × 9¼. 27471-3 Pa. $7.95

HOW WE INVENTED THE AIRPLANE: An Illustrated History, Orville Wright. Fascinating firsthand account covers early experiments, construction of planes and motors, first flights, much more. Introduction and commentary by Fred C. Kelly. 76 photographs. 96pp. 8¼ × 11. 25662-6 Pa. $8.95

THE ARTS OF THE SAILOR: Knotting, Splicing and Ropework, Hervey Garrett Smith. Indispensable shipboard reference covers tools, basic knots and useful hitches; handsewing and canvas work, more. Over 100 illustrations. Delightful reading for sea lovers. 256pp. 5⅜ × 8½. 26440-8 Pa. $7.95

FRANK LLOYD WRIGHT'S FALLINGWATER: The House and Its History, Second, Revised Edition, Donald Hoffmann. A total revision—both in text and illustrations—of the standard document on Fallingwater, the boldest, most personal architectural statement of Wright's mature years, updated with valuable new material from the recently opened Frank Lloyd Wright Archives. "Fascinating"—*The New York Times.* 116 illustrations. 128pp. 9¼ × 10¾. 27430-6 Pa. $10.95

PHOTOGRAPHIC SKETCHBOOK OF THE CIVIL WAR, Alexander Gardner. 100 photos taken on field during the Civil War. Famous shots of Manassas, Harper's Ferry, Lincoln, Richmond, slave pens, etc. 244pp. 10⅝ × 8¼.
22731-6 Pa. $9.95

FIVE ACRES AND INDEPENDENCE, Maurice G. Kains. Great back-to-the-land classic explains basics of self-sufficient farming. The one book to get. 95 illustrations. 397pp. 5⅜ × 8½.
20974-1 Pa. $7.95

SONGS OF EASTERN BIRDS, Dr. Donald J. Borror. Songs and calls of 60 species most common to eastern U.S.: warblers, woodpeckers, flycatchers, thrushes, larks, many more in high-quality recording.
Cassette and manual 99912-2 $8.95

A MODERN HERBAL, Margaret Grieve. Much the fullest, most exact, most useful compilation of herbal material. Gigantic alphabetical encyclopedia, from aconite to zedoary, gives botanical information, medical properties, folklore, economic uses, much else. Indispensable to serious reader. 161 illustrations. 888pp. 6½ × 9¼. 2-vol. set. (USO)
Vol. I: 22798-7 Pa. $9.95
Vol. II: 22799-5 Pa. $9.95

HIDDEN TREASURE MAZE BOOK, Dave Phillips. Solve 34 challenging mazes accompanied by heroic tales of adventure. Evil dragons, people-eating plants, bloodthirsty giants, many more dangerous adversaries lurk at every twist and turn. 34 mazes, stories, solutions. 48pp. 8¼ × 11.
24566-7 Pa. $2.95

LETTERS OF W. A. MOZART, Wolfgang A. Mozart. Remarkable letters show bawdy wit, humor, imagination, musical insights, contemporary musical world; includes some letters from Leopold Mozart. 276pp. 5⅜ × 8½. 22859-2 Pa. $7.95

BASIC PRINCIPLES OF CLASSICAL BALLET, Agrippina Vaganova. Great Russian theoretician, teacher explains methods for teaching classical ballet. 118 illustrations. 175pp. 5⅜ × 8½.
22036-2 Pa. $4.95

THE JUMPING FROG, Mark Twain. Revenge edition. The original story of The Celebrated Jumping Frog of Calaveras County, a hapless French translation, and Twain's hilarious "retranslation" from the French. 12 illustrations. 66pp. 5⅜ × 8½.
22686-7 Pa. $3.95

BEST REMEMBERED POEMS, Martin Gardner (ed.). The 126 poems in this superb collection of 19th- and 20th-century British and American verse range from Shelley's "To a Skylark" to the impassioned "Renascence" of Edna St. Vincent Millay and to Edward Lear's whimsical "The Owl and the Pussycat." 224pp. 5⅜ × 8½.
27165-X Pa. $4.95

COMPLETE SONNETS, William Shakespeare. Over 150 exquisite poems deal with love, friendship, the tyranny of time, beauty's evanescence, death and other themes in language of remarkable power, precision and beauty. Glossary of archaic terms. 80pp. 5³⁄₁₆ × 8¼.
26686-9 Pa. $1.00

BODIES IN A BOOKSHOP, R. T. Campbell. Challenging mystery of blackmail and murder with ingenious plot and superbly drawn characters. In the best tradition of British suspense fiction. 192pp. 5⅜ × 8½.
24720-1 Pa. $5.95

THE WIT AND HUMOR OF OSCAR WILDE, Alvin Redman (ed.). More than 1,000 ripostes, paradoxes, wisecracks: Work is the curse of the drinking classes; I can resist everything except temptation; etc. 258pp. 5⅜ × 8½. 20602-5 Pa. $5.95

SHAKESPEARE LEXICON AND QUOTATION DICTIONARY, Alexander Schmidt. Full definitions, locations, shades of meaning in every word in plays and poems. More than 50,000 exact quotations. 1,485pp. 6½ × 9¼. 2-vol. set.
Vol. I: 22726-X Pa. $16.95
Vol. 2: 22727-8 Pa. $15.95

SELECTED POEMS, Emily Dickinson. Over 100 best-known, best-loved poems by one of America's foremost poets, reprinted from authoritative early editions. No comparable edition at this price. Index of first lines. 64pp. 5³⁄₁₆ × 8¼. 26466-1 Pa. $1.00

CELEBRATED CASES OF JUDGE DEE (DEE GOONG AN), translated by Robert van Gulik. Authentic 18th-century Chinese detective novel; Dee and associates solve three interlocked cases. Led to van Gulik's own stories with same characters. Extensive introduction. 9 illustrations. 237pp. 5⅜ × 8½. 23337-5 Pa. $6.95

THE MALLEUS MALEFICARUM OF KRAMER AND SPRENGER, translated by Montague Summers. Full text of most important witchhunter's "bible," used by both Catholics and Protestants. 278pp. 6⅝ × 10. 22802-9 Pa. $11.95

SPANISH STORIES/CUENTOS ESPAÑOLES: A Dual-Language Book, Angel Flores (ed.). Unique format offers 13 great stories in Spanish by Cervantes, Borges, others. Faithful English translations on facing pages. 352pp. 5⅜ × 8½. 25399-6 Pa. $8.95

THE CHICAGO WORLD'S FAIR OF 1893: A Photographic Record, Stanley Appelbaum (ed.). 128 rare photos show 200 buildings, Beaux-Arts architecture, Midway, original Ferris Wheel, Edison's kinetoscope, more. Architectural emphasis; full text. 116pp. 8¼ × 11. 23990-X Pa. $9.95

OLD QUEENS, N.Y., IN EARLY PHOTOGRAPHS, Vincent F. Seyfried and William Asadorian. Over 160 rare photographs of Maspeth, Jamaica, Jackson Heights, and other areas. Vintage views of DeWitt Clinton mansion, 1939 World's Fair and more. Captions. 192pp. 8⅞ × 11. 26358-4 Pa. $12.95

CAPTURED BY THE INDIANS: 15 Firsthand Accounts, 1750–1870, Frederick Drimmer. Astounding true historical accounts of grisly torture, bloody conflicts, relentless pursuits, miraculous escapes and more, by people who lived to tell the tale. 384pp. 5⅜ × 8½. 24901-8 Pa. $8.95

THE WORLD'S GREAT SPEECHES, Lewis Copeland and Lawrence W. Lamm (eds.). Vast collection of 278 speeches of Greeks to 1970. Powerful and effective models; unique look at history. 842pp. 5⅜ × 8½. 20468-5 Pa. $14.95

THE BOOK OF THE SWORD, Sir Richard F. Burton. Great Victorian scholar/adventurer's eloquent, erudite history of the "queen of weapons"—from prehistory to early Roman Empire. Evolution and development of early swords, variations (sabre, broadsword, cutlass, scimitar, etc.), much more. 336pp. 6⅛ × 9¼. 25434-8 Pa. $8.95

AUTOBIOGRAPHY: The Story of My Experiments with Truth, Mohandas K. Gandhi. Boyhood, legal studies, purification, the growth of the Satyagraha (nonviolent protest) movement. Critical, inspiring work of the man responsible for the freedom of India. 480pp. 5⅜ × 8½. (USO) 24593-4 Pa. $8.95

CELTIC MYTHS AND LEGENDS, T. W. Rolleston. Masterful retelling of Irish and Welsh stories and tales. Cuchulain, King Arthur, Deirdre, the Grail, many more. First paperback edition. 58 full-page illustrations. 512pp. 5⅜ × 8½.
26507-2 Pa. $9.95

THE PRINCIPLES OF PSYCHOLOGY, William James. Famous long course complete, unabridged. Stream of thought, time perception, memory, experimental methods; great work decades ahead of its time. 94 figures. 1,391pp. 5⅜ × 8½. 2-vol. set.
Vol. I: 20381-6 Pa. $12.95
Vol. II: 20382-4 Pa. $12.95

THE WORLD AS WILL AND REPRESENTATION, Arthur Schopenhauer. Definitive English translation of Schopenhauer's life work, correcting more than 1,000 errors, omissions in earlier translations. Translated by E. F. J. Payne. Total of 1,269pp. 5⅜ × 8½. 2-vol. set. Vol. 1: 21761-2 Pa. $11.95
Vol. 2: 21762-0 Pa. $11.95

MAGIC AND MYSTERY IN TIBET, Madame Alexandra David-Neel. Experiences among lamas, magicians, sages, sorcerers, Bonpa wizards. A true psychic discovery. 32 illustrations. 321pp. 5⅜ × 8½. (USO) 22682-4 Pa. $8.95

THE EGYPTIAN BOOK OF THE DEAD, E. A. Wallis Budge. Complete reproduction of Ani's papyrus, finest ever found. Full hieroglyphic text, interlinear transliteration, word-for-word translation, smooth translation. 533pp. 6½ × 9¼.
21866-X Pa. $9.95

MATHEMATICS FOR THE NONMATHEMATICIAN, Morris Kline. Detailed, college-level treatment of mathematics in cultural and historical context, with numerous exercises. Recommended Reading Lists. Tables. Numerous figures. 641pp. 5⅜ × 8½. 24823-2 Pa. $11.95

THEORY OF WING SECTIONS: Including a Summary of Airfoil Data, Ira H. Abbott and A. E. von Doenhoff. Concise compilation of subsonic aerodynamic characteristics of NACA wing sections, plus description of theory. 350pp. of tables. 693pp. 5⅜ × 8½. 60586-8 Pa. $14.95

THE RIME OF THE ANCIENT MARINER, Gustave Doré, S. T. Coleridge. Doré's finest work; 34 plates capture moods, subtleties of poem. Flawless full-size reproductions printed on facing pages with authoritative text of poem. "Beautiful. Simply beautiful."—*Publisher's Weekly*. 77pp. 9¼ × 12. 22305-1 Pa. $6.95

NORTH AMERICAN INDIAN DESIGNS FOR ARTISTS AND CRAFTSPEOPLE, Eva Wilson. Over 360 authentic copyright-free designs adapted from Navajo blankets, Hopi pottery, Sioux buffalo hides, more. Geometrics, symbolic figures, plant and animal motifs, etc. 128pp. 8⅜ × 11. (EUK) 25341-4 Pa. $7.95

SCULPTURE: Principles and Practice, Louis Slobodkin. Step-by-step approach to clay, plaster, metals, stone; classical and modern. 253 drawings, photos. 255pp. 8⅛ × 11. 22960-2 Pa. $10.95

THE INFLUENCE OF SEA POWER UPON HISTORY, 1660–1783, A. T. Mahan. Influential classic of naval history and tactics still used as text in war colleges. First paperback edition. 4 maps. 24 battle plans. 640pp. 5⅜ × 8½.
25509-3 Pa. $12.95

THE STORY OF THE TITANIC AS TOLD BY ITS SURVIVORS, Jack Winocour (ed.). What it was really like. Panic, despair, shocking inefficiency, and a little heroism. More thrilling than any fictional account. 26 illustrations. 320pp. 5⅜ × 8½.
20610-6 Pa. $8.95

FAIRY AND FOLK TALES OF THE IRISH PEASANTRY, William Butler Yeats (ed.). Treasury of 64 tales from the twilight world of Celtic myth and legend: "The Soul Cages," "The Kildare Pooka," "King O'Toole and his Goose," many more. Introduction and Notes by W. B. Yeats. 352pp. 5⅜ × 8½.
26941-8 Pa. $8.95

BUDDHIST MAHAYANA TEXTS, E. B. Cowell and Others (eds.). Superb, accurate translations of basic documents in Mahayana Buddhism, highly important in history of religions. The Buddha-karita of Asvaghosha, Larger Sukhavativyuha, more. 448pp. 5⅜ × 8½.
25552-2 Pa. $9.95

ONE TWO THREE . . . INFINITY: Facts and Speculations of Science, George Gamow. Great physicist's fascinating, readable overview of contemporary science: number theory, relativity, fourth dimension, entropy, genes, atomic structure, much more. 128 illustrations. Index. 352pp. 5⅜ × 8½.
25664-2 Pa. $8.95

ENGINEERING IN HISTORY, Richard Shelton Kirby, et al. Broad, nontechnical survey of history's major technological advances: birth of Greek science, industrial revolution, electricity and applied science, 20th-century automation, much more. 181 illustrations. ". . . excellent . . ."—Isis. Bibliography. vii + 530pp. 5⅜ × 8¼.
26412-2 Pa. $14.95

Prices subject to change without notice.

Available at your book dealer or write for free catalog to Dept. GI, Dover Publications, Inc., 31 East 2nd St., Mineola, N.Y. 11501. Dover publishes more than 500 books each year on science, elementary and advanced mathematics, biology, music, art, literary history, social sciences and other areas.